CAMBRIDGE LIBRARY COLLECTION

Books of enduring scholarly value

Technology

The focus of this series is engineering, broadly construed. It covers technological innovation from a range of periods and cultures, but centres on the technological achievements of the industrial era in the West, particularly in the nineteenth century, as understood by their contemporaries. Infrastructure is one major focus, covering the building of railways and canals, bridges and tunnels, land drainage, the laying of submarine cables, and the construction of docks and lighthouses. Other key topics include developments in industrial and manufacturing fields such as mining technology, the production of iron and steel, the use of steam power, and chemical processes such as photography and textile dyes.

The Severn Tunnel

Costing at the time £1.8 million to complete, the Severn Tunnel was a Victorian engineering triumph, and for over a century it remained the longest rail tunnel in Britain. Construction had begun in 1873 but came to a standstill after the workings were inundated by water in 1879. An accomplished civil engineering contractor, Thomas Andrew Walker (1828–89) had worked on railways in Canada, Russia and Africa as well as on London's new underground lines; he was therefore well qualified to complete the Severn Tunnel, which was achieved in 1886. First published in 1888, Walker's first-hand account of the problematic project remains an engaging source for railway and engineering historians, and its detailed account of the ground encountered will also interest geologists. Replete with plans and maps, reissued here is the second edition of 1890, the year following Walker's death, which is likely to have been hastened by overwork.

Cambridge University Press has long been a pioneer in the reissuing of out-of-print titles from its own backlist, producing digital reprints of books that are still sought after by scholars and students but could not be reprinted economically using traditional technology. The Cambridge Library Collection extends this activity to a wider range of books which are still of importance to researchers and professionals, either for the source material they contain, or as landmarks in the history of their academic discipline.

Drawing from the world-renowned collections in the Cambridge University Library and other partner libraries, and guided by the advice of experts in each subject area, Cambridge University Press is using state-of-the-art scanning machines in its own Printing House to capture the content of each book selected for inclusion. The files are processed to give a consistently clear, crisp image, and the books finished to the high quality standard for which the Press is recognised around the world. The latest print-on-demand technology ensures that the books will remain available indefinitely, and that orders for single or multiple copies can quickly be supplied.

The Cambridge Library Collection brings back to life books of enduring scholarly value (including out-of-copyright works originally issued by other publishers) across a wide range of disciplines in the humanities and social sciences and in science and technology.

The Severn Tunnel

Its Construction and Difficulties, 1872–1887

Thomas A. Walker

CAMBRIDGE
UNIVERSITY PRESS

CAMBRIDGE UNIVERSITY PRESS

Cambridge, New York, Melbourne, Madrid, Cape Town,
Singapore, São Paolo, Delhi, Mexico City

Published in the United States of America by Cambridge University Press, New York

www.cambridge.org
Information on this title: www.cambridge.org/9781108063401

© in this compilation Cambridge University Press 2013

This edition first published 1890
This digitally printed version 2013

ISBN 978-1-108-06340-1 Paperback

THE SEVERN TUNNEL:

Its Construction and Difficulties.

1872-1887.

SIR JOHN HAWKSHAW, F.R.S.

ENGRAVED BY G J STODART FROM A PHOTOGRAPH BY BARRAUD.

London, Richard Bentley & Son, 1888.

THE

SEVERN TUNNEL:

Its Construction and Difficulties.

1872-1887.

BY

THOMAS A. WALKER.

WITH FIVE PORTRAITS ON STEEL AND UPWARDS OF FORTY SKETCHES AND PLANS,

Second Edition.

LONDON:

RICHARD BENTLEY & SON, NEW BURLINGTON STREET.

Publishers in Ordinary to Her Majesty the Queen.

1890.

DEDICATION.

THE following account of some of the difficulties met with in the construction of the Severn Tunnel, and how they were surmounted, is, by kind permission, dedicated to the Engineer-in-Chief, SIR JOHN HAWKSHAW, of whose professional skill and indomitable perseverance the Tunnel itself will ever remain a lasting monumen .

T. A. W.

June 14, 1888.

THOMAS A. WALKER.

ENGRAVED BY G.J.STODART FROM A PHOTOGRAPH BY BEATTIE.

London, Richard Bentley & Son. 1888.

PREFACE.

I was engaged for seven years in the construction of the Severn Tunnel, and during this period many visitors inspected the works.   Mr. Bentley happened on one occasion to come down from London, and, after spending the best part of a day on the works, appeared to be much interested by what he saw.   Indeed, very shortly after his return he wrote to me to ask that I would place upon record some account of the construction of the Tunnel.

I was particularly occupied at the time I unwisely acceded to this request.   Writing a book proved to me a more arduous task than a year's superintendence of the Tunnel itself.

I am reluctant to say anything about myself, but I am told it is desirable to give some brief account of my career.

Not being one of those fortunate individuals who are said to be born with a silver spoon in their mouth, I was forced to undertake responsible work in the year of the Railway Mania (1845), before I reached the age of seventeen, after a very brief course of instruction at King's College. Both in that year and 1846 I did considerable work on Parliamentary Surveys. In 1847, Mr. Brassey was good enough to give me a position on the North Staffordshire Railway, and I remained upon his staff for seven years; the last two being spent upon his great contract for the Grand Trunk Railway of Canada. For a further seven years I remained in Canada, constructing railways for the Governments of the Lower Provinces, and I returned home, after an absence of nine years, in 1861.

In 1863 I made some extensive surveys for railways in Russia. In 1864 and 1865, further surveys and explorations in Egypt and the Soudan, passing as far south as Metammeh, 100 miles north of Khartoum. On my return to England in the spring of 1865, I was offered, and accepted, the management of the construction of the Metropolitan and Metropolitan District Railways for the three firms who had jointly undertaken the contract, namely, Messrs. Peto and Betts, John Kelk, and Waring Brothers. The construction of these lines from Edgware Road to the Mansion House I carried out successfully, and they were completed on the 1st of July, 1871. Shortly after this, I, in partnership with my brother, the late Mr. Charles Walker, undertook the contract for the extension of the East London Railway from the end of Brunel's Thames Tunnel, under the London Docks, through Wapping, Shadwell, and White-chapel. Sir John Hawkshaw was the engineer-in-chief of this work, and I was fortunate in gaining his good opinion, and carried out the works, I believe, to his complete satisfaction; and it was owing to the confidence he reposed in me that he afterwards intrusted to me the still more difficult work of con-structing the Severn Tunnel.

Sub-aqueous tunnels have recently become quite the fashion. One such experience as the Severn Tunnel, with its ever-varying and strangely contorted strata, and the dangers from floods above and floods below, has been sufficient for me. One sub-aqueous tunnel is quite enough for a lifetime.

Since these pages were commenced, I have had a great pressure of work upon me. Not only have I had to carry on such large works as the Barry Dock and Railways, and the Preston Dock, but I have also been called upon to visit South America to start the work of the Government Docks at Buenos Ayres, and at home to begin the construction of the Manchester

Ship Canal. Any oversight or clerical error which may have escaped notice during the revision of the proofs will under these circumstances, I trust, receive the indulgence of the reader.

T. A. W.

INTRODUCTORY NOTE.

In the short time which has elapsed since the pub-
lication, in 1888, of 'The Severn Tunnel: its
Construction and Difficulties,' the busy career of
its able author has come to a close. Mr. Thomas
Andrew Walker died on the 25th November, 1889,
at Mount Ballan, not far from the great work which
he has so well described in these pages.

He was sixty-one years of age when he died: at a
time of life when many years of useful work might
reasonably have been hoped for from him. Yet those
who knew him well had felt, for some time before
his death, that he had taken burdens upon his
shoulders sufficient to try the health of the strongest
man, and had noticed with increasing anxiety each
succeeding symptom of failing health.

Although the record of the quantity of work
carried out by Mr. Walker does not equal that
executed by some well-known contractors—such as
Mr. Brassey, under whom he first began his career;
yet, had he lived to a greater age, there is little

doubt that he would have had few rivals even as regards quantity of work done. At the time of his death he was carrying out two contracts—the Manchester Ship Canal and the Buenos Ayres Harbour Works—the aggregate cost of which alone may be put at at least £12,000,000.

If we turn from the quantity of work done, to its character, the work carried out by Mr. Walker, under my own supervision, was such as to try the ingenuity and resources of a contractor to their utmost. Professionally, my acquaintance with him as a contractor first began in the East London Railway extension from Wapping to Shadwell. The contract for this work was taken by Mr. Walker, in conjunction with his brother Charles who died during the progress of the work. After his death Mr. Walker personally undertook the conduct of the work, and satisfactorily completed it. The railway had to be carried beneath the London Docks, and the conditions imposed by the Dock Company made that part of the work one of the very greatest difficulty. A coffer dam had to be made extending halfway across the dock, and after the tunnel (for two lines of railway) had been built within it, the coffer-dam had to be removed. After its removal, a second coffer-dam was made, within which to build the second half of the tunnel under the Dock. The difficulty of effecting the junction of the two tunnels in the centre of the Dock will be obvious to those who have any knowledge of such works.

The railway tunnel was successfully made through the Dock and beneath the Dock walls. The tunnel was also carried beneath the sugar warehouse on the north quay of the Dock; a work requiring great care, as the warehouse had basements, and the floors over the basements were entirely supported on groined brick arches. The brick piers supporting these arches were all under-pinned with concrete as far down as the foundation of the tunnel itself, after which ground was excavated between the piers; and the basement then presented a most striking appearance when illuminated by the lights of the warehouse, the groined brick arches being supported on columns 60 feet high. The brickwork of the tunnel was built round and among these concrete columns, which were afterwards removed wholly from within the tunnel, and the concrete where it passed through the brickwork was then cut out and replaced by brickwork. Other heavy under-pinning works occurred on this railway, one of which I may also mention where the tunnel passed at some depth below the brick viaduct of the Blackwall Railway.

I have spoken at some length of this work, which has not been described elsewhere, because not only did Mr. Walker display great skill in combating the difficulties met with, but many of his best men on the Severn Tunnel works had been employed on the East London Railway, and the experience gained on it was of great value in carrying out the greater undertaking of the Severn Tunnel.

Another work which tested Mr. Walker's ability was the Inner Circle completion Link from the Mansion House to Aldgate, with a branch to the East London Railway; and on that work, also, I had ample opportunity of judging his great capabilities as a contractor. How well he carried out the work of the Severn Tunnel, and the resources he showed, will be apparent to those who read this book.

In my long experience of contractors, extending over more than fifty years, I have never met with anyone surpassing Mr. Walker for despatch in carrying out works. This arose not only from his great anxiety always to fulfil his engagements, but also from the very great interest he took in all the constructive details of his work. This led him to disregard considerations of expense when difficulties were met with which had to be overcome. If more plant was required it was procured at once, if more temporary work was necessary it was ordered to be done forthwith; and so questions of loss or gain to himself never caused delay. Moreover, taking such interest as he did in his works, it followed that none but the best class of work would satisfy him.

Mr. Walker's practice of going into all the details of his works himself brought upon him an immense amount of labour over and above that which must necessarily fall to the lot of a large contractor, and it probably tended to shorten his life. When the works he had in hand were of smaller magnitude the course he pursued was practicable; but when, as at the close of his life, they had become of great extent and importance, it ceased to be possible for one brain to deal with all the details arising from them.

I know of no contractor who has displayed so much care and solicitude for the comfort and welfare of the workpeople employed by him, as Mr. Walker.

On the Severn Tunnel works, where no accommodation existed for the large number of workmen there gathered together, Mr. Walker not only built substantial and comfortable cottages, but he provided hospitals, with trained nurses, and built coffee taverns and mission halls. So, in the River Plate, at his quarries in the Banda Oriental, where some 500 men are employed, he has provided good stone cottages, medical attendance, and a mission hall, where services are held in English and Spanish.

Mr. Walker was a hard worker, never sparing himself; and that, combined with his constant consideration for those whom he employed, and a most conscientious performance of his duties, will cause his death to be regretted by all who were brought in contact with him.

JOHN HAWKSHAW.

June, 1890.

LIST OF PORTRAITS ON STEEL.

LIST OF ILLUSTRATIONS.

TELFORD'S BRIDGE, GLOUCESTER.

LONDON. RICHARD BENTLEY & SON. 1887.

THE SEVERN TUNNEL:

ITS CONSTRUCTION AND DIFFICULTIES.

CHAPTER I.

DESCRIPTION OF THE ESTUARY OF THE SEVERN, AND
THE COUNTRY IN THE IMMEDIATE NEIGHBOURHOOD
OF THE SEVERN TUNNEL.

THE River Severn, after a long course from its
source in Plynlimmon, widens out just below
Gloucester into a broad estuary, which has formed a
great obstacle to traffic passing between Bristol and
the South-West of England and South Wales.

The Severn, as a river, may be said to end at
Gloucester, at the point where the turnpike-road is
carried, by Telford's famous bridge of only 150 feet
span, over it, for almost directly below it opens into
a tidal estuary, which spreads out till, at the point
where the tunnel passes under it, it is $2\frac{1}{4}$ miles wide.
The tides in the Severn estuary are known to be
the highest in England or in Europe. They are
only surpassed in height by the tides which run up

Description of the Severn. the Petitcodiac River at the head of the Bay of Fundy, in New Brunswick.

The great rise of this tide is caused by the funnel-like shape of the estuary. The tide running round the South of Ireland, and becoming imprisoned between South Wales and the Cornish and Devonshire coasts, as the width of the channel is continually decreasing, mounts up to a great height, till it reaches, at the mouth of the Wye, a height, at spring-tides, of 50 feet above low water.

About 26 miles below Gloucester, where the river is 3,700 feet wide, it is crossed by an iron bridge constructed to carry the Severn and Wye Railway over it, and so form a connection between Lydney and the coalfields of the Forest of Dean on the west bank, and the Midland Railway between Birmingham and Bristol and the docks, at Sharpness Point, which are situated directly below the bridge on the east. The bridge was opened on the 19th October, 1879.

Before this date there had long been rivalry between the two schemes for a bridge or a tunnel.

The Great Western Railway Company had been anxious to establish a more direct route between Bristol and South Wales, and to avoid the heavy gradients of the Stroud Valley between Gloucester and Swindon. They obtained, many years ago, an Act to construct a bridge over the river near Chepstow, but the project had been abandoned; and they had finally, in 1871, adopted Mr. Charles

THE SEVERN BRIDGE.

LONDON. RICHARD BENTLEY & SON, 1867.

Maclure & C° Lith London

Richardson's plan for the tunnel under the river, which has since been carried out.

The Midland Company had in the meantime given some support to the bridge at Sharpness Point. The two works had been commenced almost simultaneously ; but when the bridge was ready for opening, in October, 1879, the only work done at the tunnel was the sinking of five shafts, and the driving of about two miles of small heading.

Among the guests invited to the luncheon by which the opening of the bridge was celebrated were Sir Daniel Gooch, the Chairman of the Great Western Railway Company, and Mr. Charles Richardson, the engineer under whose superintendence the works of the tunnel were being carried out. Sir Daniel, whose health was proposed at the luncheon, in replying, gave the company present an invitation to attend at the Severn Tunnel in about six weeks, and walk through the headings, which would then be completed. He said : 'It will be rather wet, and you had better bring your umbrellas.' Alas, he little knew how wet it was ; for Mr. Richardson, sitting near him and hearing these words, had received an intimation on his way to attend the ceremony that a great spring had been tapped on the western side of the river ; that the pumps had been overpowered by the inrush of the water, and that the whole of the work was drowned.

To Sharpness a considerable number of ships are

towed up the estuary of the Severn, and docked
at Sharpness Point, from which place there is a
ship canal, known as the Berkeley Canal, to
Gloucester.

Below Sharpness the estuary of the Severn continues to exhibit the same features—a waste of sand
at low water and a broad channel of dirty water at
high tide. The banks are generally low till we reach
Aust Cliff, where the east bank rises to about 100 feet.
This cliff shows the geological strata in a very distinct manner, the lower part being of the new red
sandstone and the upper part lias. The cliff is
famous for the number and beauty of the fossils
which are obtained from the lias beds.

On the opposite side of the river there is a small
island, on which are the ruins of the Chapel of St.
Tecla; and here was one of the ferries by which,
in the old days, general traffic between Bristol and
South Wales was conveyed across the river. This
is known as 'Old Passage;' and immediately below
the river Wye runs into the Severn, about two miles
from the town of Chepstow.

Chepstow, which contains about 3,000 inhabitants,
is a picturesque town, with part of the old walls still
remaining, as well as the ruins of the castle.

It is a favourite resort for tourists intending to
visit the Wye Valley and Tintern Abbey, the latter
being about $4\frac{1}{2}$ miles to the north-west.

The tide rises at Chepstow Bridge to the height
of 50 feet, and runs up the river Wye at high spring-

SUDBROOK CHAPEL.

Maclure & Co. Lith. London

LONDON   RICHARD BENTLEY & SON, 1887

tides for a distance of nearly 20 miles, measured along the winding course of the stream.

At the point where the Wye joins the Severn the estuary becomes at once 2 miles wide, and there is a depth of 70 feet at low water in the deep-water channel known as ' The Shoots ;' and about 2 miles below the Old Passage are the piers, built by the Bristol and South Wales Union Railway Company, where passengers travelling by the trains from Bristol to South Wales used to cross the river by steamboats. This railway and the piers are the property of the Great Western Railway Company, and, considering the difficulties encountered in the passage, a very considerable traffic was carried by the steamboats. Immediately alongside the present piers are the remains of the old wharves and roads by which the coach passengers in the old days were brought over the beach and put on board the ferry barges.

Fifty years ago this was the only route by which travellers from Bristol or Bath could reach South Wales, except by sea.

About half a mile below the ' Black Rock Pier,' on the western side of the river, on a point of land which has been saved to a great extent from the wasting influences of the sea by the hardness of the strata of which it is composed, are the remains of a Roman camp. Rather more than half the camp has been washed away by the waves, but the two sides which remain are still very perfect. Some

Description of the Severn. antiquaries have questioned whether it were a British, Roman, or even Danish camp, but a careful consideration leaves no possibility of doubt that it was a Roman camp.

The two sides that remain are at right angles to each other, and the camp was probably a square, two sides facing to the river, and the other two to the land.

In the ditch of the camp was built, in early Norman times, a small church, the ruins of which are now known as Sudbrook Chapel, Sudbrook having been a parish itself up to about 200 years ago, when, the population having almost entirely deserted it, it was joined to the neighbouring parish of Portskewett.

The point chosen for commencing the works of the Severn Tunnel is within about 100 yards of this Roman camp, on the north side of it.

To the south of the camp the inroads of the sea have made marked progress during the last century.

More than 2 miles from the shore is a small island, known as ' The Denny,' on which it is reported that, within a hundred years, a fox, followed from the mainland, was killed. It is still possible, at low water of the spring-tides, to walk to the Denny and return in the same tide, but the journey is not unattended with danger.

Opposite the Roman camp, at a distance of about 5 miles, are the new docks at Avonmouth, and the King's Road, where ships waiting to enter Portishead or Avonmouth or Bristol docks lie at anchor.

Rather more to the south, at a distance of about 7 miles, is the dark headland of Portishead, with the docks at its foot. On a clear day, a long strip of the coast in the direction of Weston-super-Mare is to be seen, and at times the island known as the 'Steep Holm,' lying below Cardiff, also.

The sea-wall which protected the meadows south-west of the camp has been entirely destroyed for a distance of many miles, and the wasting of the land still continues.

In consequence of the destruction of the sea-wall, the equinoctial spring-tides flow over a vast extent of meadow-land ; and on one occasion, as we shall afterwards have to relate, the water passed over the whole of the meadows to a depth of more than 5 feet.

From the Roman camp at Sudbrook there can still be traced the remains of a Roman road, running nearly north-west, to intersect the main road, which passed through Chepstow in the direction of Caer-leon.

At rather less than a mile from the camp is the village of Portskewett, beautifully situated, but ex-hibiting signs of neglect, with its ugly, squalid and dirty cottages and farmyards ; and above the village rises Portskewett Hill, where the mountain limestone has been upheaved.

The Roman road from the camp continues its course till it intersects the main road at the hamlet of Crick, about 1 mile to the west of which is the

present village of Caerwent, once the famous Roman
station, 'Venta Silurum.' The Roman walls remain
in fair preservation, and it is believed that when
this station was held by the Roman Legions, the
tide from the Severn flowed up to the base of the
southern wall, and that the rings to which the boats
were moored still remain.

Where these tides flowed is now a rough piece of
marsh land, through which the little river Neddern
passes to join the Severn.

The whole of the ground in the marsh is rotten,
and before the tunnel was commenced there were
enormous springs of bright clear water rising up in
several places.

At about 2 miles farther north than Caerwent, the
hills of Wentwood are met with, with 'Grey Hill'
standing in the foreground. The first spurs of the
hills fronting the valley are composed of mountain
limestone, the higher parts about Shirenewton of
the old red sandstone.

On the east side of the Severn, and for some
little distance on the western side, the new red
sandstone formation is found in nearly horizontal
beds. The first disturbance of this takes place
behind Portskewett village, where the mountain
limestone has been upheaved and the new red
formation denuded. A mile further up the same
limestone has been upheaved between Caldicot and
Caerwent, and from there to the base of the hills
the strata have been much broken, and the conse-

quence has been that all the water from the hills, both from the mountain limestone and the old red sandstone, has found subterranean channels through this broken ground, and, before the tunnel was commenced, flowed out in the valley of the Neddern, and formed the great springs which have been before mentioned.

The Neddern, rising as a small brook in the hills above Llanvair Discoed, sometimes lost the whole of its water in the dry season near the foot of the hills, bursting out again near Caerwent, at a point called by the natives ' The Whirly Holes.'

When the tunnel was being made, and a fissure was unfortunately tapped in the rock between Sudbrook camp and Portskewett village, all these underground channels poured their water into the tunnel itself, and almost every well and spring, and the little river itself for a distance of more than 5 miles from the tunnel, became dry.

The little river in its course to the sea from Caerwent passes the village of Caldicot, much the largest village in the neighbourhood, with its picturesque church, the rector of which, the Rev. E. Turberville Williams, took a most genial interest in all our works ; and just below the village are the ruins of Caldicot Castle.

Below the Castle, where the Neddern enters the estuary of the Severn, were Caldicot Wire Works, recently converted into tin-plate works, which gave employment to a number of people. The place is known as 'Caldicot Pyll.'

It is rather a puzzle to an ordinary Englishman to find every little stream where it enters the sea or a greater river, in this district, called 'Pyll ;' and again to find the same word used all round Glastonbury and applied to the drains in the marshes there. The word in the Welsh is 'Pwll,' and corresponds to the English 'Pool,' though we seem slightly to have altered the original signification.

This description of the Severn estuary and the country immediately adjoining the tunnel may be found useful in understanding the history of the work itself, and explaining many of the difficulties that were encountered when the works were being carried out.

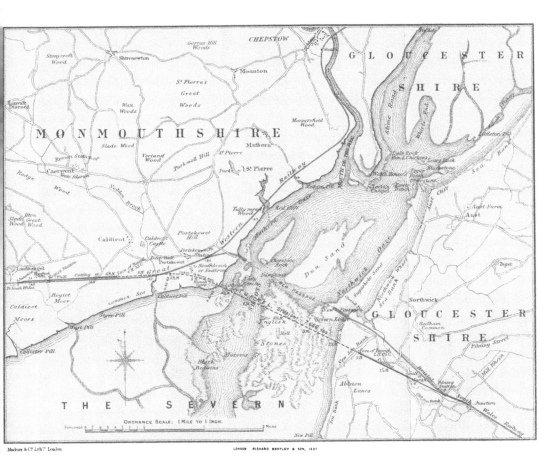

CHAPTER II.

THE EARLY HISTORY OF THE UNDERTAKING.

It has already been stated that the Great Western Railway Company had for many years been anxious to obtain a better access to South Wales.

Their main line at first ran from London to Bristol. From this, in 1838, wishing to acquire the traffic from Gloucester to London, and probably fearing that any attempt to obtain powers from Parliament to bridge the Severn below that point would be defeated, they constructed the line passing by Kemble and Stroud to Gloucester. At Gloucester the South Wales Railway joined them.

The line through the Stroud Valley has very heavy gradients and many sharp curves.

Many schemes have been set on foot to avoid these curves and gradients, and the Bristol and South Wales Union Railway, with its ferry from New Passage to Portskewett, is one of them; but this line was only a single line, with one gradient of 1 in 60, and the fatal drawback of the steamboat-ferry.

In November, 1871, Mr. Charles Richardson de-
posited plans in Parliament for the Severn Tunnel.
He soon afterwards obtained the assistance of Sir
John Hawkshaw, who agreed to act as consulting-
engineer, and the scheme being taken up by the
Great Western Railway Company, was carried
through Parliament, and an Act for the construction
of the tunnel obtained in 1872.

The Great Western Railway Company lost no
time in commencing the works, which they did early
in 1873.

In order to test the strata, they sank and lined
with brickwork, on the Monmouthshire or west side
of the Severn, a shaft 15 feet in diameter to a depth
of about 200 feet; and from this they commenced to
drive a heading eastwards under the river.

The heading had a rise from the bottom of the
shaft of 1 in 500, and was driven at the level necessary to drain the lowest point of the intended tunnel
under the deep-water channel of the river.

When the works were commenced, the parish of
Portskewett, in which they were situated, was a
purely agricultural parish, with a population of men,
women, and children, of 260. No buildings whatever existed near the site of the shaft, the nearest
being a farmhouse, known as Sudbrook or Southbrook Farm, with two cottages, and at a slightly
greater distance along the river-bank an inn, known
as the 'Black Rock Hotel,' with three or four cottages, occupied by the inspector of the steamboat

CHARLES RICHARDSON.

ENGRAVED BY W H GIBBS FROM A PHOTOGRAPH BY MAULL & FOX

London: Richard Bentley & Son: 1888

pier, and other railway servants. A footpath along Early history of the under-taking.
the bank of the river led from the nearest public
road, and from this inn to the site of the shaft. 1877.

Before commencing the works it was necessary to
make some provision for the men who would be
employed there. For this purpose land, to a small
extent, was purchased by the Great Western Railway
Company near the shaft, and upon it were built six
cottages and a small office.

A single line of railway, or tramway, was laid from
Portskewett Station to the shaft on land leased from
the tenant of Southbrook Farm. Over this the
winding engines, pumping engines, bricks, timber,
and other materials were brought to the shaft, and
six other cottages were built upon land belonging to
the Great Western Railway Company, near the
Bristol and South Wales Union Line, between
Portskewett Station and Portskewett Pier.

The progress made with so small a provision was
necessarily slow, and in August, 1877, after four and
a half years' work, all that had been done consisted
of the sinking of the one shaft, afterwards known
as the 'Old Shaft,' and the driving of about 1,600
yards of 7-feet square heading under the river. A
second shaft had been commenced, in which it was
intended to fix the permanent pumps to drain the
tunnel. This shaft in August was about half sunk,
but not lined.

At this date the Directors decided that they would
let the contract for the works, and advertisements

asking for tenders were published. Only three
tenders were received, one of which was from myself. Sir John Hawkshaw, acting as consulting
engineer, advised the Directors to accept the tender
I then made ; but, after considerable discussion, they,
being of the opinion that too great an amount had
been estimated for contingencies, decided not to let
the contract until they had further proved the ground,
and, in fact, till they had driven a heading through
the whole length of the tunnel.

They then entered into two small contracts: one
with Mr. Oliver Norris, of New Passage, to sink a
shaft on the Gloucestershire side of the river, and to
drive headings east and west from that shaft ; and
another with Mr. Rowland Brotherhood, to sink two
shafts, known as the 'Marsh Shaft' and the 'Hill
Shaft,' and to drive headings both ways from these
shafts. The Company continued to carry on the
heading under the river themselves, and they also at
a later period agreed with Mr. Norris to drive from
the original or Old Pit 7-feet headings, westwards towards the Marsh Pit, and eastwards on the formation
level of the tunnel under the river.

The Company also completed the pumping-shaft
which they had commenced in 1877. This shaft
they tubbed with iron, erected an engine-house over
it, and in this fixed two Bull-engines, each with a
50-inch cylinder and 10-feet stroke, and each working a 26-inch plunger-pump.

The iron tubbing of the shaft did not reach quite

PLAN OF SUDBROOK, 1873.

SCALE

River Severn

120 CHAINS RADIUS

To Severn Wales

SOUTHBROOK FARM

CENTRE LINE OF SEVERN TUNNEL

SUDBROOK

PARISH

PORTSKEWETT

OF

HILL HOUSE BARN

Old Holy Trinity Church in Ruins

Roman Camp Ruins

LINE OF TRAMWAY

LOCO SHED

Portskewett Junction

To Old Road to Chepstow

BLACK ROCK HOTEL

LONDON. RICHARD BENTLEY & SON. 1887.

Maclure & Co Lith London.

to the bottom, 10 feet being lined with brickwork below the tubbing.

This brickwork carried four wrought-iron girders, on which the pole-cases of the 26-inch plunger-pumps were fixed. There was a small iron door, 2 feet square in the iron tubbing of the shaft opening outwards, through which access could be obtained to a cross-heading leading to the main heading from the bottom of the Old Pit, and in the brickwork between two of the wrought-iron girders was fixed a sluice, by closing which water could be excluded from the pit, or admitted by opening it.

The girders when fixed proved too weak for the work they had to do, and an ordinary cast-iron pipe 15 inches in diameter was placed under one of the girders as a column. This pipe had, of course, a large flange at the top, which afterwards proved a serious obstacle in the way of fixing other pumps.

The valve, by which the quantity of water to be admitted to the pit was regulated, instead of being actuated by long rods brought up above the highest level of the water, was fitted with an ingenious (?) apparatus by which it was intended to shut or open it by turning on the pressure of water obtained from a spring behind the iron tubbing about 100 feet above the valve. When the shaft was full of water the action of this arrangement was most uncertain, and was the cause of much of the difficulty encountered in clearing the tunnel of water.

In the manner above described the work was proceeded with till the 18th October, 1879, at which time a considerable length of heading had been driven under the land from the three additional shafts which had been sunk, and the heading under the river had been so far advanced that only about 130 yards intervened between the heading being driven from the Gloucestershire shaft, known as the 'Sea-Wall Shaft,' and the main heading from the Old Pit on the Monmouthshire side.

None of these headings, up to the 17th October, had given any large quantity of water.

There were fixed at the Hill Pit two 15-inch plunger-pumps ; at the Marsh Pit and the Sea-Wall Pit, each two 15-inch plunger-pumps. In the Old Pit there was an 18-inch plunger-pump worked by a 41-inch Cornish beam-engine ; and in the Iron Pit adjoining were the two 26-inch plunger-pumps, each worked by a 50-inch Bull-engine.

But on the 18th October, 1879, in the heading then being driven westwards from the Old Pit, a large body of water was tapped, which, although efforts were made to dam it out by timber placed across the heading, poured into the workings in such a volume, that in twenty-four hours the whole of the workings which were in connection with the Old Pit were full up to the level of the tide-water in the river. Fortunately no lives were lost, the men being warned as they were changing shifts in the long heading, and being able to escape by the Iron Pit with only a wetting.

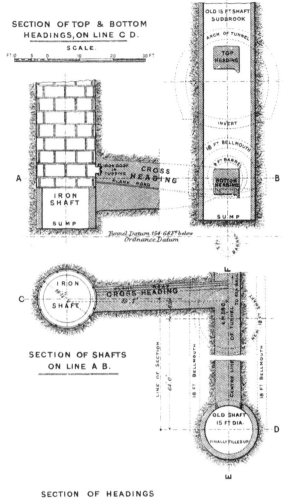

SECTIONS SHEWING RELATIVE POSITIONS OF OLD PIT, BOTH HEADINGS, AND CROSS HEADING.

SECTION OF TOP & BOTTOM HEADINGS, ON LINE C D.

SCALE.

Tunnel Datum 154·68Ft below Ordnance Datum

OLD 15 FT SHAFT SUDBROOK
ARCH OF TUNNEL
TOP HEADING
INVERT
18 FT BELLMOUTH
9 FT BARREL
BOTTOM HEADING
SUMP

A — IRON SHAFT — SUMP
CROSS HEADING — IRON DOOR TUBBING — PLANK ROAD
B

SECTION OF SHAFTS ON LINE A B.

C — IRON SHAFT
CROSS HEADING 8·1
LINE OF SECTION
18 FT BELLMOUTH
64·0
CENTRE LINE OF TUNNEL 28·6½
18 FT BELLMOUTH
F
OLD SHAFT 15 FT DIA.
FINALLY FILLED UP
D
E

SECTION OF HEADINGS ON LINE E F.

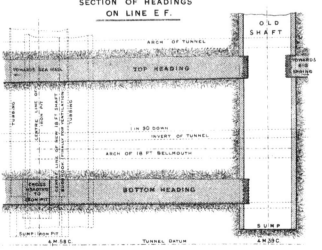

ARCH OF TUNNEL
OLD SHAFT
TOWARDS SEA WALL
TOP HEADING
TOWARDS BIG SPRING
TUBBING
CENTRE LINE OF IRON PIT
CENTRE LINE OF NEW 18 FT SHAFT
CENTRE LINE OF SUDBROOK (FINALLY FOR VENTILATION)
TUBBING
1 IN 90 DOWN
INVERT OF TUNNEL
ARCH OF 18 FT BELLMOUTH
CROSS HEADING IRON PIT
BOTTOM HEADING
SUMP
SUMP: IRON PIT
4 M 58 C.
TUNNEL DATUM
4 M 59 C.

LONDON   RICHARD BENTLEY & SON, 1887

Maclure & Co Ltd London

To understand the difficulty of their escape, it must be understood that the heading into which the water broke was 40 feet above the heading under the river, and the water falling from the westward heading made a sheer leap of 40 feet down the Old Pit and cut off all escape by it; the men had therefore to pass through the small iron door into the Iron Pit to effect their escape.

It was a melancholy result of nearly seven years' work, and no doubt those in authority no longer undervalued the contingencies of such a work, which they had thought had been over-estimated by the contractors two years before.

The Directors then called in Sir John Hawkshaw, who up to this time had been consulting engineer, and asked him to take full charge of the works as chief engineer, and carry them on as he thought best himself.

Sir John Hawkshaw agreed to take charge only on the condition that they would allow him to let the works to some one in whom he could have confidence; and on their consenting to this condition, he did me the honour to send for me, and asked me if I were still willing to enter into a contract.

Even after this irruption of the water under the land (water which it was well known was perfectly fresh and sweet), no one seemed to be fully alive to the fact that the greatest dangers and difficulties were to be found there, but everyone still thought the danger lay in the construction of the tunnel under the deepest part of the river. Sir John had to explain to me that the Directors had determined, in the first place, to complete the tunnel under what was known as the 'Shoots' (that is, the low-water channel of the river), before proceeding further with the tunnel under the land, and he asked me to give him a price for executing that part of the work only.

After a little consideration, I declined to do this, but offered to carry out the works on the tender I had made in 1877, executing the work under the 'Shoots' first, if he thought that advisable. I gave to him then all the details of the tender I had made in 1877. The directors accepted the tender, with certain modifications necessary in consequence of the lapse of time and the work that had been done, and the contract was entered into with me, and the works handed over to me on the 18th December, 1879.

Nothing could be more desolate than the appearance of the works at this time. There were, as I have stated, near the main shaft only six cottages and a small office, the necessary boiler-houses and engine-houses, a small carpenter's shop, a fitter's shop, a blacksmith's shop, and two low buildings or sheds used as cottages also. The tramway which had been originally laid to Portskewett Station had been pulled up, and in lieu of it another tramway had been laid, following (on the surface of the ground) the centre line of the tunnel itself from the Old Shaft to the Marsh Pit, and joining the Great Western Railway

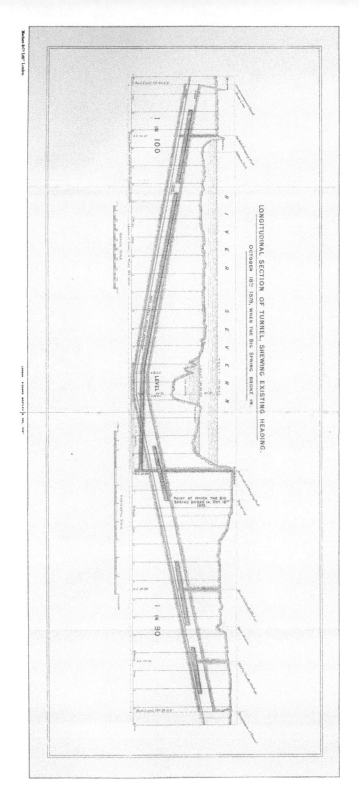

LONGITUDINAL SECTION OF TUNNEL, SHEWING EXISTING HEADING.

OCTOBER 18TH 1879, WHEN THE BIG SPRING BROKE IN.

RIVER SEVERN

POINT AT WHICH THE BIG
SPRING BROKE IN. OCT. 18TH
1879.

The material originally positioned here is too large for reproduction in this reissue. A PDF can be downloaded from the web address given on page iv of this book, by clicking on 'Resources Available'.

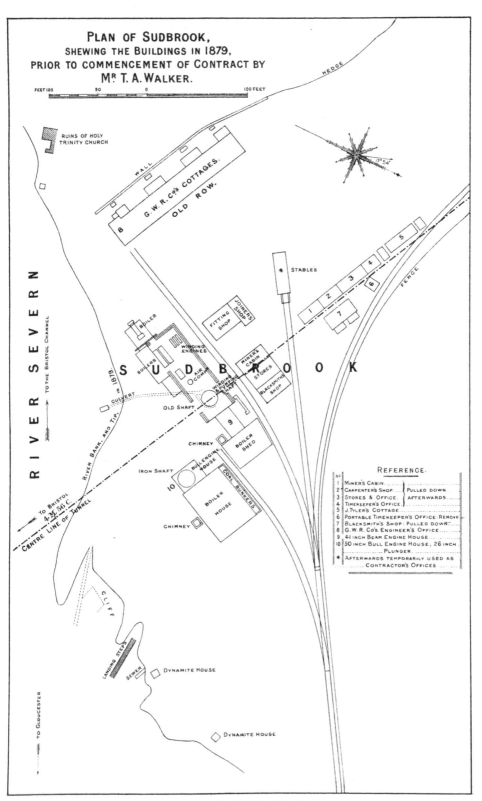

PLAN OF SUDBROOK,
SHEWING THE BUILDINGS IN 1879,
PRIOR TO COMMENCEMENT OF CONTRACT BY
MR T. A. WALKER.

FEET 100   50   0   100 FEET

RUINS OF HOLY
TRINITY CHURCH

WALL

G.W.R. Co's COTTAGES.

OLD ROW.

8

HEDGE

17° 54'

FENCE

5

4

3   6

2

1   7

STABLES

JOINERS
SHOP

FITTING
SHOP

BOILER

WINDING
ENGINES

RIVER   SEVERN

TO THE BRISTOL CHANNEL

S   U   D   B   R   O   O   K

BOILERS

AIR
COMP

MINER'S
CABIN

STORES

BLACKSMITHS
SHOP

CULVERT

IN 1879.

RIVER BANK AND TIP.

WINDING
ENGINE &
OLD SHAFT   PUMPING SHAFT

To Bristol
4 M. 56 C.

CENTRE LINE OF TUNNEL

IRON SHAFT

CHIMNEY

BULL ENGINE
HOUSE

9

BOILER
SHED

10

BOILER
HOUSE

COAL BUNKERS

CHIMNEY

CLIFF

LANDING STEPS

SEWER

DYNAMITE HOUSE

TO GLOUCESTER

DYNAMITE HOUSE

REFERENCE.

N°		
1	MINER'S CABIN.	
2	CARPENTER'S SHOP.	PULLED DOWN
3	STORES & OFFICE.	AFTERWARDS.
4	TIMEKEEPER'S OFFICE.	
5	J. TYLER'S COTTAGE.	
6	PORTABLE TIMEKEEPER'S OFFICE: REMOVD	
7	BLACKSMITH'S SHOP: PULLED DOWN.	
8	G.W.R. Co's ENGINEER'S OFFICE.	
9	41 INCH BEAM ENGINE HOUSE.	
10	50 INCH BULL ENGINE HOUSE, 26 INCH PLUNGER.	
*	AFTERWARDS TEMPORARILY USED AS CONTRACTOR'S OFFICES	

Machure & Co. Lith. London.

LONDON. RICHARD BENTLEY & SON, 1887.

a mile west of Portskewett Station at Caldicot Pyll.

The engines at the main shaft stood idle, the boilers were out of steam, most of the men who had been employed had left in search of other work, and the water in the shaft was standing up to the level of high water in the Severn.

The pumping-engines at Sea-Wall, Marsh, and Hill Pits were still at work, as the working of those pits had been in the hands of Mr. Norris and Mr. Brotherhood; but no work was being done below, orders having been given to these gentlemen to suspend their operations.

The pumps were able to keep the Hill Pit dry, but were not sufficiently powerful to drain either of the dipping headings from the Marsh or Sea-Wall Shafts while the main shaft and heading were full of water.

Immediately after the irruption of the Great Spring, and before the contract was let to me, the Great Western Railway Company, under Sir John Hawkshaw's advice, had ordered two additional pumping-engines with large pumps to contend with the extra volume of water. The one was a Cornish beam-engine with a 75-inch cylinder, intended to work either a 38-inch bucket-pump or a 37-inch plunger-pump. The other was a 70-inch cylinder beam-engine, to work two 28-inch bucket-pumps.

It may be as well here—as this is intended to be a popular account, and not written for professional readers—to state the difference between bucket and plunger pumps.

Early history
of the under-
taking.

1879.

The bucket-pump is simply the ordinary lift-pump used all over the country, with a bottom valve fixed as near as possible down to the level from which the water is to be pumped, and with large rods running down the centre of the pump attached to the bucket-valve, which works up and down the length called the working barrel, the bucket-valve opening as the pump-rods descend, and allowing the water to pass to the upper side of the bucket; while the valves of the bucket close when the engine makes the upward stroke, and the valve at the bottom opening allows the water to enter, and to follow the bucket in its ascent.

The plunger-pump is a force-pump; the pump-barrel, or rising-main, has no rods inside it, but near the bottom is connected by a casting, called the 'H-piece,' with a separate closed cast-iron cylinder known as the 'pole-case.' There are two valves, as in the bucket-pump, but they are both fixtures.

The top and bottom of the pole-case are closed, but the plunger at the lower end of the pump-rods works through a stuffing-box on the top of the pole-case, and when the rods are drawn upwards the bottom valve opens, while the top one is held closed by the column of water above it, and the atmospheric pressure drives the water through the bottom valve into the pole-case. On the return-stroke of the engine the pump-rods descending close the bottom valve, and force the water through the

'H-piece' and the top valve, up the main column of
the pump, and out to the required level.

The plunger is much the better pump for continuous work, as any slight wear there may be upon the pole itself is easily made good by packing the stuffing-box, while the faces of the buckets in the bucket-pumps are continually wearing by friction against the sides of the working barrel. (The drawings of the large pumps and engines at Sudbrook, given at the end of this volume, will show the difference between the bucket and the plunger pump.) Of course we should have preferred that all the pumps should have been plungers, but it is necessary in fixing a plunger-pump to get to the bottom of the pump to fix the pole-case as well as the bottom of the rising main securely, and after it is at work to pack the stuffing-boxes. It was impossible to put down plunger-pumps while the shafts were full of water, and it was for this reason that, in the first place, bucket-pumps were adopted and ordered.

The engines for working the pumps were of two sorts, beam-engines and Bull-engines—the latter being so called after the inventor. Both engines take steam only at one end of the cylinder.

The beam-engine takes steam only above the piston. The pressure of the steam when admitted forces down the piston, pulls down the inner end of the beam to the full extent of the stroke, raising the outer end of the beam which carries the pump-rods, and thus making the up-stroke of the pump.

Early history
of the under-
taking.

1879.

When it has reached the top, the weight of the pump-rods (assisted, if necessary, by weights placed on the inner end of the beam) makes the down-stroke without using steam.

In the 75-inch engine, working a 35-inch bucket-pump, the weight of the wrought-iron beam was nearly 23 tons, and the weight of the pump-rods was nearly 12 tons. The engine could run eleven strokes per minute, raising 376 gallons of water 173 feet high at each stroke. Steam was admitted to the cylinder at 45 lbs. pressure. The vacuum was about $13\frac{1}{2}$ lbs. Steam was cut off at less than one-third of the stroke. The engine worked best at about eight strokes per minute, giving 3,000 gallons raised in that time.

The Bull-engine takes steam only below the piston, the pump-rods being attached to and working in a direct line from the piston-rod. Steam being admitted below the piston, the pump-rods are raised by a direct lift. Their weight, which must be greater than the column of water, makes the return-stroke without steam. The Bull-engine is the most compact-looking, but does not work so economically as the beam. The heavy beam itself, being once set in motion by steam, acts as an accumulator, allowing steam to be used more expansively than is possible in the Bull-engine.

In the 50-inch Bulls, steam was admitted at 45 lbs. pressure, and cut off at one-third of the stroke. The pump-rods weighed about $5\frac{1}{2}$ tons, and the

weight of the plunger and balance-weights was
about 18 tons. The pumps worked best at eight
strokes, raising 231 gallons per stroke each, but could
be run up to thirteen strokes, per minute.

The contract contained a general description of
the work, stating that the tunnel would be 7,942
yards in length, or just over 4½ miles; that the
total length of the railway included in the contract
was 7 miles 5 furlongs; that the work under the
deepest part of the river, known as the 'Shoots,'
for a length of 660 yards, should be proceeded
with before the other work was commenced, unless
the engineer-in-chief should otherwise order. The
contract then stated that certain drawings were
signed as forming part of the contract; that the
engineer had power to substitute other drawings and
make any alteration that he thought necessary during
the progress of the works; and it was provided that
if any alterations were made, additions to or deduc-
tions from the contract sum should be made by the
engineer to compensate for these alterations.

The work that had been done by the Company
was then set forth in the contract, and the pumps
and other plant already provided or ordered by
them was also set forth in detail; and it was
arranged that payment should be made for the use
of this plant by the contractor.

The usual clauses were inserted as to the power
of the engineer over the works, as to progress and
other matters.

The Company agreed to furnish all the land required permanently for the works; the contractor
undertook to provide all that was required for temporary purposes.

The contractor was made responsible for the
setting out of the works, for the lines and levels; and
the Company, on their part, agreed to the usual
terms for payment monthly as the works proceeded.

The open cuttings at both ends of the tunnel
were to be surrounded by embankments several feet
above the level of the highest recorded tide, so that
the most extreme high tide should not, when it
flooded the meadows, run down the open cuttings
and drown the tunnel.

The tunnel was specified to be lined with brickwork, three bricks, or 2 feet 3 inches, thick; but only
half the tunnel was to have an invert. The brickwork was to be of Cattybrook vitrified or Staffordshire brindle bricks, approved by the engineer.

The mortar was to be made of one part of Portland cement and two parts of perfectly clean, sharp
sand.

The concrete, if used, was to be one part of Portland cement and five parts of sand and gravel.

There were to be a number of bridges over the
open cuttings, which were all specified in detail.

The ballast for the permanent way was to be
broken stone in the bottom, with gravel or hard slag
for upper ballast. The rails and sleepers for the
permanent way were to be furnished by the Com-

pany; the road to be a cross-sleepered road, with
steel rails weighing 86 lbs. to the lineal yard of rail.

The original amount of the contract was very considerably increased, firstly, by the whole of the tunnel having an invert; secondly, by the brickwork lining being made 3 feet thick, instead of 2 feet 3 inches, over a considerable part of the length of the tunnel; and, thirdly, from the lowering of the whole of the gradients, which will be mentioned later.

The clause in the contract which stated that the 660 yards under the 'Shoots' were to be completed before the rest of the work was proceeded with, was inserted because the Directors and the engineer still thought the chief danger in constructing the tunnel was from an irruption of the river-water; and no one supposed that there lay hidden under the land, within a quarter of a mile of the western shore, a greater difficulty and danger than any to be met with under the bed of the river. The very precautions which were taken for security under the river were the cause of magnifying the other difficulties, and there never was a clearer case of 'De Scyllâ in Charybdim.'

CHAPTER III.

THE COMMENCEMENT OF THE WORKS UNDER THE CONTRACT.

**Commence-
ment of the
works.**

———

1880.

IMMEDIATELY after the signing of the contract, arrangements were made to commence building the new engine-houses and erect the large engines which had been purchased by the Company. It was estimated that it would take rather more than six months to complete the erection of these engines and of the pumps connected with them. The work to be done was the erection of an engine-house for the 75-inch engine, with a boiler-house containing six Cornish boilers; a separate engine-house for the 70-inch engine, with a boiler-house for seven Cornish boilers of rather smaller dimensions.

The engines and pumps were made by Messrs. Harvey and Co., of Hayle in Cornwall, and they were not able to promise delivery of the machinery in less than four or five months. It was thought, therefore, that the works would continue in their then state of desolation up to midsummer, 1880, except for the work to be done in erecting these houses and the machinery.

On obtaining, however, full particulars of the manner in which the headings had been worked from the Old Pit, especially the heading running westwards into which the Great Spring had burst, I thought that it would be quite possible to stop the mouth of the heading and the flow of water from the Great Spring by lowering into the shaft two shields made to fit, as nearly as possible, to the interior of the shaft, and strutted across the shaft in a secure manner. The difficulty was that the mouth of the heading was then 140 feet below the surface of the water ; but, having obtained the sanction of Sir John Hawkshaw to make the attempt, my plans were made, and as early as the 6th of January, eighteen days after the contract was signed, operations were commenced.

According to the best information we were able to obtain from those who had been in charge of the works, the quantity of water running into the heading under the river was not more than 2,000 gallons per minute, and to deal with this we had, in the Iron Pit, two 26-inch plunger-pumps, each capable of lifting 2,500 gallons per minute, and in the Old Pit an 18-inch plunger-pump, capable of lifting 1,200 gallons per minute ; so that if we could by any means succeed in stopping back the Great Spring, we had ample power to clear the rest of the works of water.

The construction of the shield and the method of securing it in the shaft are shown upon the accompanying plate.

Commence-
ment of the
works.

1880.

The shields were of oak, made to fit as nearly as possible to the interior of the shaft. One shield, when put together, was lowered to cover the entrance to the western heading; then the second in like manner lowered to cover the entrance to the eastern heading; and when lowered to their proper position, heavy struts, also of oak, were to be fixed between them and wedged up tight.

In addition to this the faces of the shields, where they were to come in contact with the brickwork, were padded with soft material soaked in tar.

The only difficulty about the operation was that the shields must be lowered by signals received from divers, who must be down the shaft at the level of the headings, and that the struts must be fixed and wedged up also by divers at the same level.

The depth of water from the shield to the surface being about 140 feet, the pressure upon the divers was so great that very few men were able to bear it at all, and no man could do work requiring great physical exertion under that pressure. In order to reduce the pressure to some extent, Sir John Hawkshaw consented to my starting the pumps and lowering the water in the shaft 50 feet. The three pumps were accordingly started on the 6th January, and the lowering of the shields commenced immediately afterwards.

On the 10th January, No. 2 26-inch pump broke down; the top valve, not being properly secured and

SHIELD IN OLD SHAFT SUDBROOK WITH DOORS.

OPENING IN SHIELD

W.C. Plate 3×5/8

W.C. Plate 3×5/8

DETAILS OF DOOR

SCALES.

Inches 12 0 1 2 3 4 5 6 7 8 9 10 11 12 Feet

Doors opened by rope worked above to help diver.

TOP HEADING

OLD SHAFT

SUDBROOK

Plates for Covering Shields

TOP HEADING

TOP HEADING

OLD SHAFT

SUDBROOK

TOP HEADING

Inches 12 6 0

LONDON RICHARD BENTLEY & SON, 1867.

Maclure & Co. Lith. London.

held down, jumped out of its seat and would not Commencement of the
works.
work.

1880.

Arrangements were made at once to lower in the
Old Pit a 15-inch pump to help to hold the water
down till the shields were fixed.

The shield to close the western heading was
made with two doors opening as flaps into the heading; the size of the opening of each door was
2 feet 3 inches by 1 foot 6 inches.

On the 14th we found, to our surprise, that the top
valve of No. 2 pump had, from vibration or some
other cause, gone back to its seat again, and we
re-started the pump; but on the 16th, to add to
our troubles, No. 1 pump, which had hitherto been
working well, broke down utterly, and the pump was
useless.

It was impossible to attempt to repair this pump
under water, and our only hope then was that we
might be able, by closing the door and the sluice at
the bottom of the Iron Pit, to empty the pit and then
repair it. If the shaft had been properly constructed
there should have been no difficulty in doing this,
the shaft being tubbed with iron to within 10 feet of
the bottom. Unfortunately the lower 10 feet was
lined with brickwork only 18 inches thick. This
was cut away on the north and south sides of the
shaft to receive the iron girders and the sluice, and
the brickwork had not been properly made good
round these. We therefore found every effort that
we made to clear the pit of water futile. More water

came in through the brickwork and the sluice when it was shut than one 26-inch pump would lift, *i.e.*, more than 2,500 gallons per minute.

No. 2 pump, however, continuing to work fairly well, we continued lowering the shields, and, with occasional stoppages for repairs, we completed the fixing of the shields by the 24th January. We at once found that the shield in the western heading leaked badly all round.

The flap-doors were opened, and a large number of bags full of Portland cement was passed through by the divers, and built up inside the shield for the purpose of stopping any leakage of water.

On the 29th January, this cement being all placed behind the shield, the pumps were stopped to allow it to set properly before strain was brought upon it.

A 15-inch pump was added to the 18-inch in the Old Pit, worked by the 41-inch beam-engine, and on the 5th February the pumps were again started, and by the 7th the water was lowered in the Iron Pit 72 feet.

A diver was then able to examine No. 1 26-inch pump, which had broken down; and he found that the H-piece was broken.

Many plucky attempts were made by the divers, especially by Lambert, to ascertain where the water came in at the bottom of the Iron Pit, and whether anything could be done to stop the leakage, so that the one pump might empty the pit, and permit us to repair the broken H-piece.

On one occasion, on the 9th February, the suction of the pump drew Lambert so fast against the perforated wind-bore or suction-piece, that it required three men upon a rope to pull him away.

We were compelled, after this experience, to give up all thoughts of repairing the broken H-piece for the present.

On the 16th February, Lambert, the diver, reported that he found the door in the tubbing of the Iron Pit was not properly closed, the rubber with which it was faced being turned up on one side, and allowing a considerable quantity of water to leak through.

The pumps were stopped on the 16th February, to allow the water to rise in the pit, and Lambert opened the door, put the rubber straight, and closed it again. This was all done in two hours, and the pumps started again. On the 20th we succeeded in reducing the water in the Iron Pit to 20 feet above the bottom of the shaft, and for a short time we had strong hopes of being able to clear the pit of water and repair the broken pump; but, on the 21st, another accident happened to No. 2 pump—one of the tappets being bent. In two hours a new tappet was put on, and the pump started again.

I determined to remove two more 15-inch pumps from the Hill Pit, and fix them in the Old Pit to increase the power there; but on the 2nd March the top valve of No. 2 pump jumped out of its seat again.

Commence-
ment of the
works.

1880.

We then got an extra valve made to drop down inside the rising-main of the pump. It was attached to wooden pump-rods 6 inches square. When lowered nearly to the top of the original valve, it was held down in position by the rods, and the pump started on the 31st. We found that it did fair duty, but after working four days it was evident that the shield would not hold back the water sufficiently to allow the pumps we then had to deal with it, and the attempt to pump out the shaft was abandoned till the new 75-inch engine with the 38-inch bucket-pump was ready for work.

In the meantime Sir John Hawkshaw had decided in January to lower the gradient of the tunnel at the 'Shoots' 15 feet, and to keep the gradient eastwards towards Bristol parallel to the old gradient at a depth of 15 feet below it, by this means preserving the gradient of 1 in 100 against the load. From the 'Shoots' westward the gradient was altered to 1 in 90, so that the lowering at the Old Pit at Sudbrook was 12 feet 7 inches; at the Marsh Pit, 7 feet 9 inches; and at the Hill Pit, 4 feet 9 inches; and the gradient ran out into the old levels in the open cutting on the Monmouthshire side.

Having decided to lower the gradient, and feeling thus greater security under the river, Sir John decided to allow me to commence the work at other points.

A shaft, 18 feet in diameter, was commenced at Sudbrook, on the centre line, directly opposite the

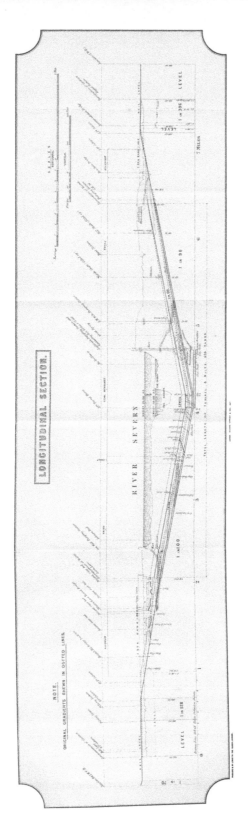

The material originally positioned here is too large for reproduction in this reissue. A PDF can be downloaded from the web address given on page iv of this book, by clicking on 'Resources Available'.

iron shaft, on the 11th February, and arrangements Commencement of the works.
were made for sinking a pumping-shaft on the
Gloucestershire side at Sea-Wall, which was to be 1880.
45 feet to the south of the existing pit, and kept as
a separate pumping-pit.

It was also arranged to sink two shafts at a point
about 26 chains from the Sudbrook shafts westwards, to commence the work to the west of the
point where the Great Spring had broken in, and it
was at that time thought possible that we might,
after these shafts were sunk, arrange to pump all
the water from the Great Spring at that point, and
keep the rest of the work dry.

The shaft on the Gloucestershire side, known as
the Sea-Wall Shaft, was to be sunk 27 feet below
the old levels to allow for the lowering of the
gradient 15 feet, and for sufficient sump-room to
hold the water for the pumps. The sinking of the
shaft was commenced on the 18th March, and was
proceeded with, with a little difficulty, till the level
of the heading was reached. A cross-heading was
driven from the new pit to the old one, 10 feet
above the bottom of the old pumps, and what water
was found in the new pit was allowed to flow through
this to the pumps.

The sinking below that level was much more
difficult. There were difficulties with boilers, and
difficulties with pumps, and it was not till the
17th July that the shaft was finally completed, and
the pumps fixed. The brickwork of the shaft had

been completed on the 7th, and ten days had been occupied in removing the pumps from the old pit to the new.

A heading had been driven eastwards before the contract was let to me, for a distance of 1,020 feet from the Sea-Wall Shaft. Some time in the summer of 1879, the roof at the end of this heading broke through the marl into a bed of gravel. There had been a great run of gravel and water, and the heading was partly filled. A head-wall had been built across the heading, 240 feet from the shaft, with a door. The door had then been closed, and the water from the gravel shut out. Before the new shaft was commenced, this door was opened, and the heading east of it examined. It was found to be in a bad state, supported by very small timbers, and not at all safe. Some extra timber was put in, and the door, which had been fixed by the Company in this heading, was again closed.

Another cross-heading was driven to suit the new gradients after the pumps were fixed at Sea-Wall, and a commencement was made at lowering the heading westwards to the altered levels.

To secure the ground, a small piece of the tunnel was commenced in a peculiar way, which was afterwards largely adopted.

The lowering of the gradient to so great an extent as 15 feet made the existing headings useless for drainage purposes, and it was decided to put in the arch of the tunnel down to springing

Cross Section of Tunnel,

SHEWING MODE OF CONSTRUCTION

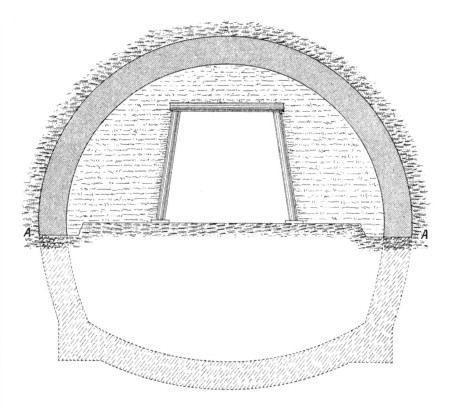

*The Arch was built to the line **A.A.***
*The lower part shewn in dotted lines was built later and the
side walls underpinned up to the Arch*

SCALE

INCHES 12 0 1 2 3 4 5 6 7 8 9 10 11 12 13 14 15 FEET

LONDON. RICHARD BENTLEY & SON, 1887.

only, and after the headings had been lowered, to
put in the invert and underpin up to the arch. This
was carried out through more than a mile of the
work, and everywhere successfully.

The new winding-pit at Sudbrook was commenced on the 11th February, and was sunk to a
depth of 30 feet by the 8th March. The shaft was
21 feet in diameter outside the brickwork, and
18 feet inside. It was impossible to sink this shaft
more than 40 feet till the water was pumped out of
the adjoining works. A bore-hole was started on
the 20th April to allow the water to drain out of the
pit into the heading below, when the new pumping-
plant should have cleared the other works of water.

The two new pits at 5 miles 4 chains, known as
pumping and winding shafts, were commenced on
the 23rd February; and on the same day a long
culvert, to convey the water out to the edge of the
river, was also commenced. These two shafts
reached a depth of 30 feet on the 8th March. On
the 21st April the winding-pit was 63 feet deep, and
the pumping-pit 65 feet. On the 2nd June the
winding-pit reached a bed of very hard conglomerate rock at a depth of 96 feet. This rock,
which was known to exist, and had been found in
the Sudbrook Pit at nearly the same level, had been
nowhere met with in a greater thickness than 9 feet,
and in many places it was known to be only 3 feet
thick. In this unfortunate shaft the thickness proved
to be 26 feet. The rock was very jointy, and full of

Commencement of the works.

1880.

fissures, yielding immense quantities of water, the water spouting through the fissures under a head of about 100 feet; and it was not till the 18th July that the shaft reached the bottom side of the bed, and entered the fire-clay shale, which proved to be perfectly dry. The bottom of the tunnel was reached at the end of July, and profiting by the experience which we had obtained at Sea-Wall, we drove a cross-heading 4 feet 3 inches below the level of the invert of the tunnel in the direction of the pumping-pit. From this heading we sunk the bottom part of the pumping-pit in shale, which was perfectly dry till we reached within a few inches of the bottom, where a small spring was met with, which evidently was under a pressure of at least 100 feet.

By putting in a small iron pipe to allow this water to rise into the cross-heading, we were able to complete the brickwork at the bottom of the pumping-pit before we allowed the water from the conglomerate above to come down upon it. When this brickwork was completed and the cement properly set, a bore-hole was put in down the centre of the shaft, and the water allowed to flow into the bottom of the pumping-pit and through the cross-heading to pumps fixed in the line of the tunnel. The shaft was then completed, both as to sinking and brickwork. The pumping during the sinking of these pits had been done by a variety of pumps, but finally by a 15-inch bucket-pump.

While the shafts were being sunk a large engine-house had been built opposite the pumping-pit, and

in it had been fixed a 70-inch Cornish beam-engine. A horizontal engine had also been fixed at the pump-ing-pit, with two 15-inch bucket-pumps, and these pumps started work on the 1st November.

The fixing of the larger pumps in the pit, which were 28-inch bucket-pumps, was also pushed for-ward, and the engine with the one 28-inch bucket-pump was started on the 15th November, and con-tinued steadily to work till the 18th December ; by which date the second 28-inch pump had been com-pleted, and the engine was then started working the two 28-inch pumps.

On the 31st May permission was given to start the tunnel-works on the Sea-Wall side of the river, and the mining for the shaft-length was commenced on the 22nd June.

As there was a prospect now of getting a strong force to work at various points, it became necessary to arrange dwelling-houses for the men.

On the 9th March I arranged with the owner of the land, Mr. C. E. Lewis, of St. Pierre, to take on lease for a period of six years a plot of land near the main shaft at Sudbrook, on which to erect cottages. I had also for some time been experimenting with the various clays and marls in the neighbourhood, with a view to making bricks, and, the experiments proving successful, I decided early in the year to construct a brickyard of considerable extent near the 5 miles 4 chains pit. On account of the season it was necessary to provide for drying the bricks, as they were made under cover, and we commenced the arrangement of the brickyard and the erection of the brick-drying shed on the 23rd of February.

A plan is given showing the state of the brick-yard and the houses that had been built at the end of 1880.

The bricks that were made from the marsh clay, with a small addition of sandy marl, proved to be of excellent quality, and the buildings, after six years, are as perfect as when first erected.

When the shaft at 5 miles 4 chains was sunk through the conglomerate, just above the level of the springing of the arch of the tunnel, I noticed that the material which was raised from the shaft was very similar to that used in the Cattybrook Brick-yard for making the vitrified bricks. It at once struck me that the heading into which the water had broken between Sudbrook and 5 miles 4 chains must be wholly in this fire-clay shale.

This was stoutly denied by the men who had been employed upon the works, but it proved to be the case, and I determined at once to endeavour to make a considerable quantity of vitrified bricks from this fire-clay shale upon the ground. This led to a large increase of the brick-making plant, and finally to the erection of eight Staffordshire kilns, with a drying-shed 100 feet by 150 feet, and a very heavy crushing-mill (similar to those used in Staffordshire

PLAN SHEWING BUILDINGS &c. AT SUDBROOK & 5 M. 4 CHS. DECR. 1880.

SCALE

100 50 0   1   2   3   4   5   6   7   8   900 FEET

REFERENCES

1 Contractor's Offices.
2 Fitting Shop.
3 Stores.
4 Stables.
5 Winding Engine & Air
  Compressor Shed.
6 41 Beam Engine House.
7 Lancashire Boiler House.
8 (Old) Museum Hall.
9 Private House - the Villas
10 Mr. I. Talbots House
11 Naptha Shed.
12 Stables & Coachhouse.
13 70 Beam Engine House.
14 Cornish Boiler House.
15 75 Beam Engine House.
16 80 Bull do do
17 Old Winding & Pumping Pit.
18 New do Pit.
19 Iron Pit.
20 Shafts at 5 M 4 C
21 Workmans Wooden Huts

These References also apply to the
Plans shewing Building &c. at
Sudbrook in 1882-4.

RIVER SEVERN

PARISH OF PORTSKEWETT

PORTSKEWETT

SUDBROOK (PUMPING STATION)

Remains of Roman Camp

Chapel Ruins

Edge of Top

CAMP ROAD

CENTRE LINE OF RAILWAY   126 CHAINS RADIUS

5 MILES   SHAFT BOTTOMS   5 M 4 C

5 M. 4 C. BRICKYARD

to South Wales

Maclure & Co. Lithr London.

LONDON   RICHARD   BENTLEY   &   SON,   1887.

for the blue bricks), to crush the shale before it went

to the brick-making machine.

While these works were going on, the erection of the large engine-house for the 75-inch engine had been steadily progressing. The brickwork being completed, the erection of the engine itself was commenced on the 24th April.

The lowering of the 38-inch pump in the Iron Pit was commenced on the 4th June. On the 16th June this pump fouled a piece of timber in the pit, and the diver being sent down to explore, reported that there was a heavy piece of oak lashed across the pit which prevented the pump from being lowered.

On the 17th June the one pump we had still fit for work in this pit was started to relieve the pressure for the divers to commence removing this timber. It was then found that a timber-stage extended over the bottom of the pit above the sump.

Again the divers had to descend to remove this stage. By the 25th the stage was removed and the pump lowered, when we found that it was resting on a flange of the 15-inch pipe which had been placed as a column under one of the girders carrying the 26-inch plunger-pumps. Lambert again examined the bottom of the pit, and found that the flange of this column projected nearly 6 inches from the side of the girder.

The size of the pump had been fixed by Sir John Hawkshaw, upon information received from those who had been in charge of the work for the Company when the shaft was sunk, and the size fixed (38 inches) was just as large as would pass between the girder and the brickwork in the side of the shaft. The projection of the flange of this column made it impossible to lower the pump into the sump of the shaft.

Attempts were made to remove the column sideways, but the diver was unable to do so. It was suggested that the upper flange should be cut or broken off. This was also beyond the power of the diver under such a depth of water, and almost disheartened by these repeated difficulties, it was determined to start the pump, and try to lower the water sufficiently to repair the broken H-piece of No. 1 26-inch plunger-pump.

Everything being ready, pumping was commenced with the 75-inch beam-engine and the 38-inch pump, and No. 2 Bull-engine with the 26-inch pump, at 7 a.m. on the 2nd July, and in $8\frac{1}{2}$ hours the water was lowered to within a few feet of the bottom stage; when, at 3.30 p.m , the 38-inch pump suddenly broke all to pieces, and in a few hours the shaft was again full of water.

The breakage of a pump of such large dimensions was of itself sufficient to terrify all those who were near it. Suddenly the heavy pump-rods, losing the resistance of the water, ran out with a crash, the outer end of the beam (which weighed nearly 23 tons) striking with immense force the catch-

wings, provided in case of such accidents. Before
the man could stop the engine it came in again with
a thundering blow, and then one of the men who
had been watching at the bottom of the pit ran up,
with terror in his face, to say that the great pump
had burst near the bottom, and that a piece more
than 18 inches across had been driven out of the
working-barrel close by him as he stood on the
staging. It was useless to work the other pump, so
it was stopped, and in a few hours the works at the
main shaft were again in the position they had been
seven months before—full of water to the level of
the tide.

At first no one could understand the cause of this
accident; but without wasting time we proceeded
again to pull out the 38-inch pump.

The rising-main, being of wrought-iron, was found
to be perfect as when it had been lowered; but
when we reached the working-barrel, which was of
cast-iron, it was found to be split from end to end,
and a large piece broken out of the one side, very
much as the man had described it. The valve-piece
below the working-barrel was also split, and the
valve was missing; this latter was found in the
suction-piece or wind-bore.

On examination of the pump when pulled out, we
found that the bottom valve had had no valve-seat
on which to rest, but was let into a taper in the cast-
iron valve-piece; the taper being 16 inches long,
and only ⅜ths of an inch larger at the top than at

Commence-
ment of the
works.
——
1880.

the bottom. The cause of the failure of the pump
was then seen. The valves, both bottom valve and
top valve, were of a new patent construction, called
'Hat-band Valves.' They consisted of a series of
steps bored with a large number of $\frac{1}{2}$-inch circular
holes, and each step covered with a band of rubber
$\frac{3}{4}$-inch thick; the principle being that the pressure of
the water through these holes should expand the
rubber, and find an escape for itself as the engine
made its stroke, and close again over the holes as
soon as the pressure was relieved.

It became manifest that these rubber-bands had
offered a very great resistance to the passage of the
water; and when the water was lowered to within
10 feet of the bottom of the shaft, the column of
water standing nearly 180 feet in the rising-main
above the bottom valve, the moment the bucket-
valve began to make its downward stroke, the water
not escaping freely through the bands of rubber,
increased the already enormous pressure upon the
taper-seating of the bottom valve, split the valve-
case, and drove the bottom valve into the wind-
bore. The engine thus released, ran out with
great force; the column of water in the rising-
main following. Unfortunately the bucket-rods were
attached to the bucket without key or nut, but by a
taper in the end of the rod, identical with the taper
in the bottom valve below.

The shock of the water on the return-stroke of
the engine tore the rod through this taper, splitting

the working-barrel, and driving out the large piece Commence ment of the works.
spoken of before; and the pump was a wreck.

1880.

All these matters were, of course, carefully in-quired into on the spot by Sir John Hawkshaw and his assistants; and, seeing at once that the accident was the result of a defect in the pump which they had undertaken to furnish to me for the work, he ordered the broken parts to be replaced by others.

The large pump was all taken out of the shaft by July 15. To appreciate the labour entailed, it must be understood that the pump-rods alone were balks of timber 45 feet long and 15 inches square, with heavy iron mountings; then the rising-main, con-sisting of 9-feet lengths of 40-inch wrought-iron pipes, had all to be lifted out again, unjointed and laid on one side, before we could get to the very bottom pieces which were broken.

New four-beat valves were ordered by the Great Western Railway Company from Messrs. Harvey and Co., of Hayle, with a new valve-piece and working-barrel; and a new suction piece or wind-bore, made oval instead of round, so that it might be lowered in the space between the lining of the shaft and the projecting flange of the pipe on which the girder rested. In ordering these care was taken to see that the bottom valve had a good level seat-ing in the valve-case, and that the pump-rods were properly attached to the bottom of the bucket.

The makers were pressed to use all despatch in furnishing the new work, and by dint of great exertions the pump was again ready for starting on the 12th October.

A new H-piece had also been provided for the 26-inch plunger attached to the 50-inch Bull-engine in the Iron Pit, and spare valve-pieces and valves for both the 26-inch pumps.

The new working-barrel valves and the valve-piece for the large pump were made 35 inches in diameter instead of 38 inches, and the pump was from this time spoken of as the 35-inch pump.

CHAPTER IV.

THE FINAL STRUGGLE.

THE 35-inch pump and the 75-inch engine being ready for work on the morning of the 14th October, No. 2 Bull-engine with the 26-inch pump being also in a condition to do about two-thirds duty with the drop-clack in it, which had been put down the rising-main, and the 41-inch engine with the 18-inch plunger-pump being able to do a little, though possibly not more than quarter duty, these three pumps were started at eleven o'clock on the morning of the 14th, for the final struggle with the water.

As nearly as we could tell, the sluice in the tubbing and the iron door above it were both closed. The water stood at tide-level, which was 38 feet 6 inches below the surface of the ground. The stage near the bottom of the Iron Pit covering the sump was about 190 feet below the surface.

To test the large pump, it was started running only four strokes per minute. This was increased to seven strokes per minute on the 16th, eight strokes on the 17th, nine strokes on the 18th, and gradually up to ten and a half strokes; the greatest speed at which we allowed this large pump to run being eleven strokes per minute.

After an hour's pumping the water in the Iron Pit was lowered nearly 30 feet, and there was still sufficient leakage going on through the closed sluice and door to lower the water outside the Iron Pit (that is, the water which was in the heading and the other pit alongside) 6 feet. In six hours from the time of starting the pumps the water was down 88 feet in the Iron Pit, and 9 feet throughout the whole of the works.

After twenty-four hours' pumping the water was down 121 feet in the Iron Pit, and 24 feet throughout all the workings, and four and a half hours later the Iron Pit was clear down to the stage over the sump, the water throughout the workings standing at a depth of 24 feet lower than it had been before we commenced pumping.

As soon as the stage was reached we took out the broken H-piece and lowered the new one, which had been provided to repair No. 1 26-inch pump. This pump was ready to work at midnight on the 16th. No. 1 was started and No. 2 stopped; the drop-valve was drawn out of the rising-main and the top valve of No. 2 pump put in its right place and properly secured. The two pumps were started together on the 19th, but another stoppage of six hours was necessary on that day to repair a broken joint in the valve door-piece of No. 1. On the

26th another stoppage was necessary with this pump for ten hours to repair a joint which had blown in the H-piece, but the pumping was continued steadily with fair results. On the 19th we had opened the door partially to take more water from the works, and on the 20th the water stood at 54 feet from the surface of the ground in both the pits. For the next seven days the pumping continued without incident, the lowest point to which we were able to reduce the water being 128 feet from the surface in the Iron Pit and 126 feet from the surface through the rest of the workings.

It being then evident that the power we had at command was barely more than sufficient to 'hold' the water, I decided to fix two additional 15-inch pumps in the Old Pit, and to add an 11-inch pump to the 41-inch beam-engine ; and at the same time I decided partially to close the door in the tubbing and to hold it by a wedge which could be withdrawn at any time, so that if anything further went wrong with the pumps in the Iron Pit we could close the door by withdrawing the wedge, and so make whatever repairs were necessary.

In order to fix the door in the iron tubbing in the way in which I wished to do it, it was necessary to let the water again rise to a level in both pits, so that on the 28th October, after these arrangements were made, the water was only 65 feet from the

surface of the ground in both pits. Pumping was
continued steadily, and by the 31st the depth from
the surface in the Iron Pit was 155 feet, and throughout the workings 150 feet. On the 2nd November
the water was down 161 feet in the Iron Pit, and
154 feet 6 inches in the other parts of the workings.

I then determined to endeavour to shut an iron
door in a head-wall which had been built by the Company in the long heading under the river about 1,000
feet from the bottom of the shaft, which the men in
the panic on the first breaking in of the water had
left open. The diver to do this work would only
work under a head of 30 feet of water. He would
have to walk up the heading 1,000 feet from the
bottom of the shaft, drawing his air-hose after him,
and when he reached the door he would have to go
behind the door, which opened inwards, and shut
down a flap-valve upon an 18-inch pipe ; come back
through the door, pull up two rails of the tramway,
close the door after him, and then screw down a
12-inch sluice-valve which was on another pipe on
the north side of the door, when all communication
with the further part of the heading under the river
would be completely cut off.

As it was impossible for one diver to drag so long
a length of hose as 1,000 feet after him up the
heading, three divers were engaged. One stood at
the bottom of the shaft to pass the hose, which was
floating hose, round the bend from the shaft into the
heading ; the two others then started up the heading

for a distance of 500 feet, where one remained to
pull forward the hose and feed it to the leading
diver.

The leading man, in whom I had thorough con-fidence, was named Lambert.

He started on his perilous journey armed with only a short iron bar, and carefully groped his way in total darkness over the *débris* which strewed the bottom of the heading, past upturned skips, tools, and lumps of rock, which had been left in the panic of 1879, until he reached within 100 feet from the door, when he found it was impossible to drag the air-hose after him, as it rose to the top of the head-ing, and its friction against the rock and the head-trees offered greater resistance than he could over-come. He, however, would not give up without an effort, and he pluckily sat down and drew some of the hose to him and then started on again, but after one or two vain efforts he found it impossible to proceed, and was obliged to return to the shaft defeated.

About this time I had heard of a diving-dress, patented by a Mr. Fleuss, by the use of which the diver was able to dispense entirely with the use of the air-hose, by carrying in a knapsack on his back a supply of compressed oxygen gas, which he was enabled to feed to his helmet as required.

After Lambert's failure to reach the door on the 3rd November, I telegraphed for Fleuss to bring his patent dress and try if he could do the work.

On the 4th he arrived, full of confidence in the success of his attempt.

All the instructions which could be given to him were given, and on the 5th Lambert and he descended into the heading, Lambert with the ordinary dress and the air-hose to start Fleuss fairly up the heading, and to encourage him. After three attempts on the 5th November, it became evident that Fleuss had not sufficient practice as a diver, or confidence in himself, to go so far up the heading; with some difficulty I persuaded Lambert to put on Fleuss's dress and try how he could work in it. After spending half an hour under water in this dress, Lambert returned fully satisfied, and undertook, with a little more practice, to make another attempt to get to the door, and he started to do so on the 8th.

Knowing the obstacles he would have to meet on his way, it was not without considerable anxiety that I saw Lambert start, for he would have to climb over the skips and other things before mentioned in total darkness, and I had to give him many cautions to be careful that the knapsack, on which he depended for air, should not strike the roof of the heading, or any of the timbers, and fracture the small copper pipe which led air from his knapsack to his helmet.

On the afternoon of the 8th, Lambert succeeded in reaching the door. He pulled up one of the rails and removed it, but having then been absent some time, and feeling no doubt nervous from the novelty of the experiment he was making, he returned again to the shaft without shutting the door.

Still full of confidence, he started again on the 10th, and reached the door again in safety, went through, and let down the flap-valve, pulled up the other rail, and closed the door.

He then screwed round the rod of the sluice-valve the number of turns he had been told it would take to shut it, and returned safely and in triumph to the shaft.

On this journey he was absent one hour and twenty minutes, but he showed no sign of exhaustion on his arrival at the surface.

The water at this time stood 174 feet from the surface in the Iron Pit, and about 6 feet higher in the rest of the workings.

How anxiously we watched the floats which told us the level of the water, and how great were our disappointment and annoyance when we found that it still continued to go down at the rate of only about 3 inches an hour, and even at high tide in the river to stand stationary for some hours.

The 18-inch plunger-pump in the Old Pit had been badly out of repair during the whole of this pumping. Two or three attempts had been made by Lambert to pack the stuffing-box under water in his diver's dress. The first time he did it, it only stood for half an hour, the second time for four or five hours, and at last we had to give up this attempt to repair the 18-inch pump as hopeless.

By 5 p.m. on the 11th November the water was down 184 feet from the surface, and the stuffing-box of the 18-inch plunger being then above water, we managed to pack it, though the men in doing so were working under a perfect cataract of water falling from the upper heading behind the shield.

The men were so sanguine now, that the principal foreman issued an invitation to the Company's inspector, Mr. Jackson, and others, to walk up the long heading with him on the 12th. On that day the water was all out of the Iron Pit, but there was a depth of 5 feet still in the heading outside it. We began then to open the door in the tubbing wider; but when it was full open, and the water in the Iron Pit held to the level of the stage, there was still between 3 and 4 feet in the heading.

Just when we felt assured of success, on the 13th, No. 1 26-inch pump had to be stopped for repairs. Two days later No. 2 had to be stopped to pack the piston of the engine, and on the 19th the door-piece of No. 2 pump split; but having a new one ready, we only stopped the pump two hours.

From the 19th November to the 1st December, No. 1 pump was under repair, and the water had again risen to 80 feet from the surface.

When repairing No. 1 pump, we took out the double-beat valves with which it had worked up to that time, and replaced them with four-beat valves, which worked much more smoothly.

At 6. p.m. on the 1st December the whole of the

pumps were started again, and the water was lowered as follows :

					Iron Pit.	Rest of the Workings.
Depth from surface—2nd Dec., at 12 noon					127 feet.	124 feet.
,,	,,	2nd ,,	,, 6 p.m.		135 ,,	132 ,,
,,	,,	3rd ,,	,, 12 p.m.		150 ,,	148 ,,
,,	,,	4th ,,	,, 12 p.m.		162 ,,	
,,	,,	5th ,,	,, 12 p.m.		174 ,,	

and on the 6th December the water was all out of the Iron Pit, and there was only two feet of water in the heading outside it.

On the 7th the foreman of the Cornish pumps, James Richards, was able about mid-day to walk up the heading to the door which the diver Lambert had shut, and then he found the cause of our disappointment at not gaining upon the water as soon as Lambert had succeeded in shutting the door.

The rails were properly pulled up and removed, and the door was properly closed. The flap-valve on the pipe on the south side of the door was also shut, but the sluice-valve on the other side had a left-handed screw, and the valve must have been closed when Lambert reached it ; and when he had given it the right number of turns to close the valve, instead of closing it he had opened it to its full width. Richards at once screwed down the valve, the effect was felt immediately, and the pumps were then slowed down.

Several other members of the staff took advantage of the news brought by Richards, and went down at once to explore the headings.

No one but those who have been engaged in such a struggle can imagine the delight of all hands at the victory which it had taken us nearly twelve months to win.

On the 13th December, the principal foreman of miners, Joseph Talbot, had the doors in the shields over the western heading opened, and explored the heading for a distance of nearly 600 feet ; and on the 14th Mr. J. Clarke Hawkshaw and myself, accompanied by Talbot, went through these doors and made a second exploration of the heading. We found a stream of water 7 feet wide and about a foot deep flowing down the heading; and certainly it was a novel experience to pass through the doors in the shield and get into the heading. We were all clothed in divers' dresses, with 'sou'-westers' instead of helmets ; and standing on the stage by the door, over the sill of which about 10 or 12 inches of water was running, we had to put one leg through and sit down in the water while we gathered the other leg after us, and came into a standing position in about 3 feet 6 inches of water at the back of the door. From the farthest point we could reach, we could see the broken timbers where the water had first burst in. A great quantity of *débris* had been brought in by the spring, and at about 600 feet from the door this filled the heading to a depth of between 3 and 4 feet, so that if we had wished to go farther we must have gone on all-fours. The air, too, proved bad, and lights were with difficulty kept burning ; so we

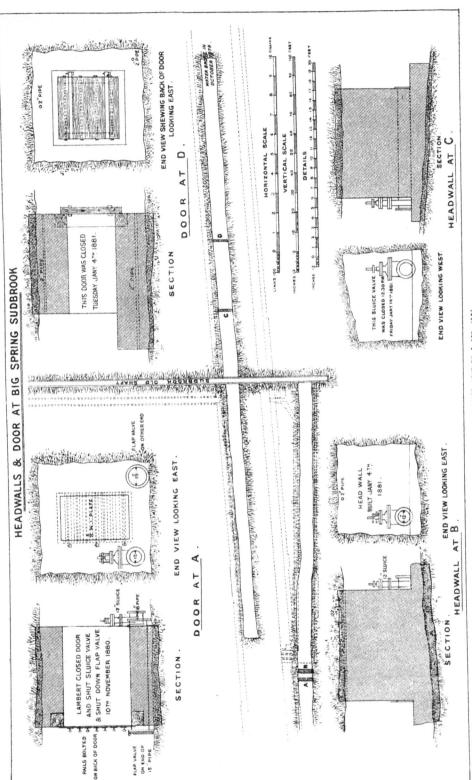

HEADWALLS & DOOR AT BIG SPRING SUDBROOK

END VIEW SHEWING BACK OF DOOR
LOOKING EAST.

O 2" PIPE

SECTION

DOOR AT D.

THIS DOOR WAS CLOSED
TUESDAY JANY 4TH 1881.

SECTION

HEADWALL AT C.

END VIEW LOOKING WEST.

THIS SLUICE VALVE
WAS CLOSED 12.30 P.M.
FRIDAY JANY 14TH 1881.

HORIZONTAL SCALE

VERTICAL SCALE

DETAILS

WATER BROKE IN
OCTOBER 1879

SUDBROOK OLD SHAFT

END VIEW LOOKING EAST.

DOOR AT A.

SECTION

FLAP VALVE
ON OTHER END

S.W. PLATE

LAMBERT CLOSED DOOR
AND SHUT SLUICE VALVE
& SHUT DOWN FLAP VALVE.
10TH NOVEMBER 1880.

RAILS BOLTED
ON BACK OF DOOR

FLAP VALVE
ON END OF
15 PIPE

12" SLUICE

END VIEW LOOKING EAST.

HEADWALL AT B.

O 2" PIPE

HEAD WALL
BUILT JANY 4TH
1881.

SECTION HEAD WALL AT B.

12" SLUICE

London. RICHARD BENTLEY & SON, 1887.

Maclure & C? Lith? London.

selected a place where the rock seemed sound for building a head-wall to stop back the water from the spring, which we decided to undertake at once.

To execute this work it was necessary to make two dams across the heading with good clay, and between the two to carry the water in wooden troughs about 3 feet square. Under the troughs a chase was cut well into the bottom of the heading; chases were also cut on both sides and in the roof, and in these a strong head wall of brickwork, in cement 3 feet thick, was built with a door-frame of 12-inch timbers, and hooks to hang the door. The 3-ft. troughs passed through the door-frame, and when the brickwork was completed the shoots were removed and the door hung, but left open.

The bricks, cement, and timber had to be taken up the heading on a rough raft pulled backwards and forwards by a rope; and, before any of these things could be done, an air-pipe had to be laid and a supply of compressed air provided for the men.

The brickwork was finished, and the door hung by about the end of the year 1880; and after allowing time for the brickwork to set, the door was closed on the 4th January, 1881, and the water from the spring entirely shut out from the works. This head-wall was 469 feet west of the centre of the Old Pit, and for more than two years from the time the door was shut we had no more trouble from the spring itself.

Thus ended the year 1880, the first year of the
works.

It was well for us that we had not known all the
difficulties we were to encounter when we entered
on the work, almost with a light heart, on the
January previous.

In addition to the work in the tunnel considerable
progress had been made in providing dwellings
for the men, and other things for their accommodation. On the first plot of leased land six large
houses had been built, which were each capable
of holding two married couples and about twelve
lodgers. Six smaller houses to accommodate a
married couple and six or eight lodgers had been
erected, as well as small houses for a married couple
and two or three lodgers or children, and a number
of semi-detached houses of a better class had been
provided for foremen. A very good house, situated
close to the shaft and the bank of the river, had been
built for the principal foreman, Joseph Talbot, and
was occupied by him and his family.

The brickyard had been thoroughly started, and
the large crushing-mill for crushing the hard shale
for making blue bricks was nearly completed.

A mission hall had been built, and arrangements
made for a supply of preachers, principally from the
Evangelization Society, and services were held on
Sundays and Wednesdays. A day-school for the
children had also been opened, and was in full working order, as well as a Sunday-school, which was

attended before the end of the year by forty
children.

A good driving-road had been made from the shaft to the nearest public road, near the Black Rock Hotel, and encouragement had been given to the various tradesmen in Chepstow and Caldicot; and from this time to the completion of the works the men were as well supplied with all the necessaries of life as if they had lived in an old town, tradesmen and farmers' carts of all descriptions calling at all the houses nearly every day.

Other land for houses had been leased, and also land for a second road leading in the direction of the villages of Portskewett and Caldicot. Arrangements had been made with the Great Western Railway Company to allow a timber bridge for this road to be thrown over their South Wales line, and the erection of the bridge was in progress.

The heading between the door, which had been shut by Lambert, and the bottom of the shaft being rendered useless as a drainage heading by the lowering of the gradient under the river, it had been intended that after the works were completed it should be filled up with rough masonry or concrete; but, on a careful examination of the heading itself, it was found that in some places it was in a very dangerous condition. For part of its length the heading was driven in the Pennant sandstone; but it also passed through beds of coal-shale, and, being very lightly timbered, was in a very bad state, and for four years it was necessary to use it as the only means of access to the works.

It was therefore arranged by Sir John Hawkshaw that, instead of ultimately filling up this heading, it should be at once lined with brickwork, being first enlarged so as to leave a circular adit, 9 feet in diameter, inside the brickwork. This it was proposed should be used, when the tunnel was completed, as an auxiliary passage for ventilation, leading to a fan which should exhaust air from the tunnel.

CHAPTER V.

GREAT SNOW-STORM——STRIKE AMONG THE MINERS——
RIVER SEVERN BREAKS IN, ON THE SEA-WALL
SIDE.

THE year 1881 opened under brighter auspices than its predecessor, the water from under the river being entirely excluded by the head-wall and door which had been built by the Company, and on the 4th January the water from the big spring also being shut out by a head-wall and door, as before stated. For double security we proceeded to build, inside each of these, a second head-wall; the one in the western heading being about half-way between the first one and the shaft, and the other one as close as we could place it to the head-wall under the river.

By building these second head-walls, all the small leakage which had come through the first was stopped out from the works.

The work of enlarging the existing heading, to line it with 18 inches of brickwork, and leave a 9-ft. barrel, was commenced; and it was well that all these precautions were taken, for a new danger

Progress of the work.

1881.

confronted us, which we had not, and could not have, foreseen.

Of course, the consumption of coal by the pumping-engines was very considerable; but as we were in communication by railway with the South Wales line of the Great Western Railway, and so with the coal-pits, we had never thought it necessary to provide any large stock of coal, when, on the 18th January, 1881, the great snow-storm, which will long be remembered in England, came upon us without warning.

I was returning to the works from London on that day, and left Paddington by the quick train, known as the 'Zulu,' at three o'clock. The storm was then raging furiously, but we got through to Swindon, an hour late, only to remain there snowed up all night. When the train reached Portskewett at one o'clock the next day, eighteen hours late, we found all the roads blocked with snow, and our branch railway, from the Great Western to the works, for most of its length blocked to a depth of between 3 and 4 feet. The goods traffic of the main line was entirely disorganized, and it was not till the 21st that we were able to obtain a supply of coal direct from the coal-pits.

We had been reduced to all kinds of expedients, begging and borrowing all the coal we could in the neighbourhood, and finally cutting up timber to keep the pumping-engines going. We were, however, able to do so; but it was not till the 28th that the severe frost which followed the snow-storm broke up, and enabled us to go on as before with the works.

All building operations on the top were, of course, stopped during this period.

The road and temporary bridge over the Great Western Railway were completed in February, and the making of vitrified bricks from the shale excavated in the tunnel was commenced in April.

The new shaft at Sudbrook, afterwards called the 'New Winding Shaft,' was pushed forward now that the water was out of the headings; and at the bottom of this shaft the 9-ft. heading was enlarged to an 18-ft. tunnel, to give way for extra lines of way for the skips, and to allow of the turning and handling of heavy timbers, which had to be lowered down the shaft for bars or sills.

The 9-ft. barrel-drain was completed to the head-wall under the river in April. Two feet above the bottom of the drain it was planked over, and on the planking a double road for skips was laid, the gauge of each road being 1 foot 9 inches. As soon as this was completed, and the road laid up to the first head-wall, that wall was cut away, and the sluice in the second head-wall opened gradually till all the water was drawn out of the heading under the river, when the door was opened and the heading was explored. On the 9th May I went up this heading with Joseph Talbot. Generally, it was in a pretty good state, because the ground was very good, but there was not timber sufficient to hold weight if there

Progress of
the work.

1881.
should be any; and at about a mile and three-quarters from the bottom of the Old Pit we found the roof had fallen in for a considerable distance, and then that a great fall had come from the roof, and the whole heading was stopped.

The air-pipes which had been laid up this heading were choked by the fall of the dirt, so that the air was bad, and lights would not burn, and we had to return from our exploration, walking something like a mile in the dark before we could obtain fresh lights.

I at once made arrangements to send men up to secure those parts, where the fall had taken place, with timber. Before it was possible for them to carry lights, it was necessary to punch holes in the air-pipe at short intervals to obtain a supply of air.

It was nearly an hour's work for men to go up, pushing trollies with timber, from the shaft to where the fall had taken place. I therefore arranged for a few men to go in, taking timber with them and food, so that they might remain at the end of the heading for the whole shift of ten hours.

The men who had been working for six years for the Great Western Railway Company, before the contract was let to me, had always felt a grudge against me, probably because they had had easier times under the old *régime.* Under the Company they had to work nominally eight hour shifts, going to work at six in the morning, firing a round of shot in the face of the heading, loading up the material which had been brought down by the blasting, and

then coming out. On this system three sets of men
worked in each twenty-four hours, changing shifts at
six in the morning, two in the afternoon, and ten at
night. Each shift was allowed half an hour in the
middle of the shift for a hasty meal.

I had insisted on the men working ten-hour shifts,
and, during the shift, coming twice to bank for their
meals.

All the work having to be done by blasting, it
was dangerous for the men to return to the faces
immediately after the shots had been fired, because
at that time dynamite was used, the fumes of which
are dangerous.

On the ten-hours system it was arranged that
nearly all the blasting was done just before the men
came out of the tunnel to their meals. The air was
then clear by the time they went down again. The
men commenced to work at six o'clock, and worked
three hours. From nine to ten they came out of
the tunnel, and had breakfast ; going down again at
ten, they worked till one o'clock ; from one to two
they came out to dinner, and from two till six they
worked again to complete the ten-hour shift.

When the Great Spring broke in and drowned the
tunnel in October, 1879, the distance from the shaft
to the face of the heading under the river was
10,100 feet, or more than a mile and three-quarters.
All the skips were brought out by men pushing
them that distance, the men being known as 'runners

out.' These 'runners out' had very short lengths to push the skips, and the cost, when the Company was working the heading, was more than ten times what it could have been done for by ponies, or perhaps twenty times what it could have been done for by proper hauling-engines.

There had been a bad spirit among the men from the time I had taken possession of the works. I believe they had wished that I should fail in pumping the water out of the tunnel. I am not quite sure that they had not wilfully caused some of the difficulties that had occurred ; and now that the works were opened throughout, and there was a prospect of making better progress, they determined to make a stand, and either force me to abandon the work altogether, or to yield to their demands. Their discontent first showed itself by their jeering at the men who took their meals with them up the long heading, asking them why they did not get tin hats made to carry their dinners in ; and then by assaulting in the darkness, or when they could meet with them alone, men I had brought from a tunnel I had just finished at Dover.

At last, on Saturday the 21st May, a notice appeared, written in chalk, at the top of the main shaft : ' I hope the ———— bond will break, and kill any man that goes down to work.'

The men gathered round the pit, but refused to go below.

It was Saturday morning, and the pay of the

night-gangs would commence between eleven and
twelve o'clock.

I may as well state here that on Saturday the men only worked seven hours, but were paid for ten. They went down, as on other days, at six; came up to breakfast from nine to ten, and the first, or day-shift, finished at two. The second, or night-shift, commencing at two, worked only till ten, making seven hours, with one off for a meal; and the works, except the pumping, were not carried on at all from 10 p.m. on Saturday till 6 a.m. on Monday.

After refusing to commence the shift they went off to the nearest public-house, came back primed with drink, and gathered in front of the pay-office grumbling; but they never came to me or the fore-man and stated any grievance or asked for any concession. They simply determined to make trouble and stop the works if they could. I was in the office at the time, so I went down into the middle of them, and said :

'Now, what do you fellows want ?'

No answer.

'Now, tell me what you want, and don't stop hanging about here.'

Then one of them said :

'We wants the eight-hour shifts.'

I said: 'My good men, you will never get that, if you stop here for a hundred years. There is a train at two o'clock, and if you don't make haste and get your money you will lose your train.　You had better get your money as soon as you can, and go.'

The men looked very sheepish, went to the pay-office and got their money, and the works were absolutely deserted for the following four days.

This strike, as I have said, occurred on Saturday, and the next night (Sunday) the great timber pier at the Black Rock, where the ferry steamers landed their passengers, was burnt down, and there were not wanting people to say that it had been burnt by the men on strike ; but, in my judgment, they were innocent of this　There had been a long period of dry weather, the timber in the pier was very old, and, above the level of high water, very dry ; and I think some pleasure-seeker on Sunday had probably thrown away a fusee after lighting his pipe, and there being a high wind at the time, the fire had spread quickly.　It probably arose either from this or from the careless raking out of the fire from the boiler of the engine which was used in lifting the luggage of passengers from the steamer at the end of the pier.

It was a good thing for the works that this strike occurred when it did, for it cleared away a number of bad characters who had gathered on the works ; and from this time to the completion of the contract there was hardly any trouble with the men, and I think there was a thoroughly good feeling between employer and employed.

The works were all secure below, and, the water

PORTSKEWETT PIER

(WITH THE WORKS — SUDBROOK IN THE BACKGROUND)

LONDON RICHARD BENTLEY & SON, 1887

from both the long heading and the Great Spring

having been shut out by the head walls, one 26-inch
pump was able to pump all the water, and the other
26-inch and the 35-inch were in reserve.

In order to bring a little pressure to bear upon
the men on strike, I ordered all the carpenters,
blacksmiths, and other men employed on the surface
to be stopped on Tuesday morning. This was a
blow they had not expected, and on Thursday, the
26th May, some of the men came back and asked
for work. On Friday a larger number returned,
and in a few days the works appeared to be going
on as before, with this great advantage, that a number of bad characters had been got rid of.

The ten-hour shifts were worked on the west side
of the river from this time to the termination of the
work without difficulty. After the men returned to
work, the slip at the end of the long heading was
thoroughly secured, the heading cleaned up to the
face, and we commenced to drive the 130 yards that
intervened between it and the heading coming from
the east.

The fall in the roof proved to be a very serious
one. For a length of 50 feet it had fallen down
from a height of 20 feet above the heading, and
though we afterwards secured the brickwork safely
through this length, we were always troubled at this
point with water, which found its way into the work;
for, by the breaking of the strata, fissures no doubt
extended at this point quite up to the river-bed.

It has been mentioned before that the Great Western Railway Company, before the contract was let to me, had not only commenced the heading westwards from the Old Pit, into which the Great Spring broke, but had also commenced a heading eastwards, on the formation level of the tunnel. The lower or long heading was driven from the Old Pit at such a level as to drain the lowest part of the tunnel under the 'Shoots' on the original gradients. This other heading, which was directly over it, and about 40 feet above it, was driven on the gradient of the tunnel, descending 1 in 100 towards the river.

As the water was lowered, this was the first part of the works we were able to explore. At a very short distance from the shaft it passed through coal-shale, and there we found that the timber had not been sufficiently strong to support the pressure of the shale. The timbers were broken, and the heading was filled up with the *débris*. We cleared this heading out and descended to the farthest point to which it had been driven, which was 864 feet from the centre of the Old Pit. We found that in order to drain the heading several bore-holes had been put down from this heading into the long heading, and, as soon as we had secured the timbering through the coal-shale, we broke up a shaft from the long heading at the extreme end of the upper one to give the men a second means of escape, should any further accident occur.

As soon as the water was out, and the lining of

the bottom heading commenced, we also commenced

the tunnel proper in the upper heading, though the lowering had made this heading neither a top nor a bottom heading, but placed it some little distance above the centre of the tunnel.

At Sea-Wall Shaft we had commenced the tunnel itself, and the length of brick arch turned at this side of the river on the 1st January, 1881, was 93 feet.

We were also driving a bottom heading at the new levels, *i.e.*, about 19 feet 6 inches below the old heading.

At a distance of 262 feet westwards of the Sea-Wall Shaft, we commenced on December 20th, 1880, a break-up to enable us to begin the tunnel proper at a second point. This was called 'Sea-Wall Break-up No. 1.' At a further distance of 200 feet from 'No. 1 Break-up' we commenced, on January 10, 1881, 'Break-up No. 2;' and at a still further distance of 161 feet, on February 6th, 1881, 'Break-up No. 3.'

At the end of April, 1881, the break-up length of brickwork was completed in No. 3 Break-up, and two lengths were turned, one at each end of it; the bricklayers were at work, and had nearly completed the second length east of the break-up, when suddenly the water burst in from the roof of the tunnel, and drove the bricklayers for a time away from their scaffold. This was a more serious difficulty than any we had hitherto met with, as it was salt water that came into the tunnel, and it was manifest that the river had to some extent broken into us.

At low water, when there remained in a pool at this point, called ' The Salmon Pool,' only about 3 feet of water, we completed the length of brick-work in the best way we could, stopping back the water with litter; and at the same time sent a number of men out on the foreshore of the river to see if they could find any holes by which the water had found its way into the tunnel. As that part of the river-bed was never actually dry at low water, and as it was impossible to indicate with any precision to the men the spot they were to search, a consider-able time was spent fruitlessly in endeavouring to find the hole. At last, by making a number of men join hands and walk through the water, the hole was found ; one of the men suddenly popping down out of sight, and being pulled out by those who had hold of his hands at either side. Having found the hole, it was only a work of time to secure it on the top ; but as there was a risk that the water running through continuously might seriously enlarge it, the pumps at the Sea-Wall side were stopped, and the water allowed to rise in the tunnel to the same height as the tide outside. In the meantime clay-puddle had been prepared, and, as soon as it was ready a schooner was loaded with it and taken out at high water to the point where the hole had been discovered; at low water the hole was effectually stopped by layers of loose clay and clay in bags alternately, with a considerable heap of clay in bags

laid over the top for the sake of weight. When this
was completed, the pumps were started again, and
the clay was found to have effectually stopped the
water.

At each end of the lengths of brickwork, where
the water had broken in, a 6-ft. length of tunnel was
immediately commenced, 4 feet thick; the ordinary
work being only 2 feet 3 inches.

These lengths were taken out as rapidly as possible,
and brickwork in cement built in them, thus forming,
in the chases which had been cut, a ring of brickwork nearly double the ordinary thickness, to stop
the water from travelling along the back of the
work. This proved entirely successful, and though
we had once or twice to repair the clay puddle on
the top and replace the bags, we had no further
trouble with the Salmon Pool.

Later on in the work we had a similar experience
with a small lake, known as 'The English Lake;'
but there we never had the same volume of water
rushing into the tunnel, and had no necessity for
stopping the pumps.

It was a most fortunate thing that this bursting in
of the water from the Salmon Pool occurred before
the long heading was completed; for had we broken
through from the long heading on the west side
of the river into the heading on the east side, it
would have been impossible to stop the pumps and
let the water rise in the tunnel till such time as

we could have built a wall across the heading to prevent the water flowing westwards ; and as this would have taken several days to do, the rushing in of the water might have done irretrievable damage.

When the men returned to work after the strike, at the end of May, as many as could work in doing it were sent up the heading from Sudbrook to secure the slip which had taken place at the east end ; but before commencing to drive the heading and complete the link between the east and west sides of the river, I found it necessary to make a better arrangement for getting out the skips than the expensive method which had been followed by the Company. The men were therefore started over a long length of the heading to slightly deepen the one side of it, and to put in, at a height of about 2 feet from the bottom, a decking of 3-inch plank-ing, where it was possible to do so. In other places to build a dry stone wall to raise the skip-road above the water-level. As the heading was only 7 feet high originally, it was necessary to blow down part of the top to increase the height, as I determined not to work on this long 'lead' till I could get ponies to do the hauling. It was necessary also to put in heavier rails than had been originally used. These had been bridge-pattern rails, about 18 lbs. to the yard, quite unsuitable for any heavy traffic.

They were replaced with rails 42 lbs. to the yard,

and a good road, with proper drainage, made throughout the whole length of 1¾ miles.

A system of sixty Swan electric lights was also established at the end next the shaft, and 20 candle-power lamps were fixed at distances of 22 yards apart, from the bottom of the shaft to where the first-break-up was to be formed, for building the tunnel under the 'Shoots.' The long heading was then driven through to join the eastern side of the river, and the junction was effected at 10 p.m. on the 26th September, 1881.

Up to the time of making this junction the only system of ventilation available for us was by conveying compressed air from compressors at Sudbrook through the whole length of 1¾ miles.

Three mains were laid for this air—one 2-inch main for the men driving the heading, and two 3-inch mains for supplying compressed air to work the rock-drills in the break-ups under the 'Shoots.'

We were using, for blasting, 'tonite,' made by the Cotton Powder Company, which I had selected as giving off less noxious fumes than any other explosive, except the highly-washed gun-cotton; but in the end of July two men, who had been working in the heading at the Marsh Pit, died of inflammation of the lungs, after a very short illness. The opinion of the doctors in charge was that the men died from inflammation of the lungs contracted from exposure, the headings being very wet and hot, and the men careless; but, as usual, some people were found to say that their deaths were due to bad ventilation and the fumes of dynamite.

We had not used dynamite in that work at all; but the death of these men made me very anxious to adopt the best possible method of ventilation, and Mr. Wales, the Government Inspector of Mines in South Wales, was good enough to pay a visit to the tunnel to inquire into the whole matter, and to give me his advice; in consequence of which I purchased a Guibal Fan, 18 feet in diameter and 7 feet wide, which was fixed at the top of the New Pit at Sudbrook.

The whole of the head-gearing of this pit was boarded in with close boarding and felted, and the entrance to it was obtained by two pairs of folding-doors, so that when the fan was at work it exhausted the air from the tunnel up the shaft, and a good current of air was set up through the whole of the workings; the fresh air being drawn in at the Sea-Wall Shaft ($2\frac{1}{4}$ miles from where the fan was fixed) as soon as the long heading was completed under the river.

The new shaft at Sudbrook was also fitted with two large iron cages, each large enough to contain four of the skips loaded with rock, or forty men, or two horses or ponies; and by using these large cages the quantity of rock, etc., drawn up from the tunnel was increased, and the time occupied by the men in changing shifts very considerably shortened.

The fan gave perfect satisfaction during the period of more than four years that it was working; and the same system is adopted for the permanent ventilation of the tunnel; a 40-ft. Guibal Fan having ultimately taken the place of the 18-ft. fan which was used while the works were in progress.

At 5 miles 4 chains the heading was driven east till we reached a distance of only 26 feet from the point where the Great Spring had broken into the heading, coming westwards from Sudbrook.

The heading was almost entirely in the fire-clay shale, and was perfectly dry. The full-sized tunnel was proceeded with and pushed forward over the same length. At the same time the heading was driven westwards from the 5 miles 4 chains pit, and the full-sized tunnel commenced from break-ups at various points, as shown on the section.

At a distance of 800 feet from the shaft the heading passed from the fire-clay into the conglomerate rock, in which the driving was most difficult, both on account of the hardness of the rock and the enormous quantity of water met with. Isolated as this pit was, and short as the lengths of the headings in communication with it were, it required a 70-inch Cornish beam-engine, with two 28-inch pumps, to keep the workings free from water.

The brickyard which had been established was close to the top of this pit, and the crushing-mill for crushing the shale was connected with the head-gearing of the pit; so that the shale from the tunnel, which was brought up in skips, was run straight to the crushing-rollers, and, within half an hour from the time the shale was got in the tunnel, it was made into bricks, and the bricks placed upon the floor of the drying-shed to be dried for the kiln.

CHAPTER VI.

TUNNELLING.

As it is impossible to write an account of this description without continually using more or less technical terms, it will be as well here to state briefly the methods by which tunnels are constructed, and so to explain many of the terms which are unavoidably used.

Generally tunnels are adopted where it is necessary to construct a railway or a canal through a hill at a greater depth than 60 feet. The comparative cost of open cutting, or tunnelling, may be said to decide whether a tunnel should be made or not; but if a tunnel is to be adopted, an engineer of experience in these matters will select a line of route and a line of levels which will bring the tunnel into the most favourable strata for its construction.

Of course, in dealing with the Severn, a tunnel under the river, or a bridge over it, could have been constructed to complete the connection between Bristol and South Wales.

If a tunnel were adopted, the only open question

would be, on what line and at what level it should
be driven to pass through the most favourable
ground, and encounter the least risk. To determine these questions, it was necessary to put down
borings and ascertain the nature of the strata. The
geology of the district was carefully studied by Mr.
Richardson, borings were taken at many points, and
the strata were supposed to lie horizontally. But in
the execution of the work they were found to be
much contorted, as will be seen by the longitudinal
section of the tunnel.

The tunnel was adopted by the Great Western
Railway Company in preference to a bridge, and the
length of 4½ miles, which was originally fixed for
tunnelling, was the length between the points at
which the gradients rose, to 60 feet from the surface.

Tunnelling may be in rock, or in very soft or loose
strata.

Tunnels may be required to be executed with
great speed; or if they are short, speed may be no
object. The consideration of speed will determine, to
a large extent, how the tunnel should be carried out.

Forty years ago there were continual discussions
as to whether the best method of constructing a
tunnel of any length was by driving a top heading
or a bottom heading. If the tunnel is short, and in
solid rock, no heading may be necessary.

In the long tunnels through the Alps, which
generally required no lining, a top heading was
driven by one gang of men, widened out by other

gangs, and lowered to the required level by other **Progress of the work.**

1881.
gangs.

But generally in tunnelling it is necessary to drive a heading—first to ensure the correctness of the line of the work ; and, secondly, to attain sufficient speed. And without doubt a bottom heading is the best for this purpose.   If the tunnel is without an invert, the heading should be driven on the level known as ' formation level ;' that is, 2 feet below rails.   If the tunnel has an invert, the heading should be driven at the level of the top of the invert, as, if it were driven at the bottom of the invert, the brickwork of each length would block the heading.

Driving the heading at the level of the top of the invert necessitates, where many lengths are going on at the same time, bridging over each length as the invert is taken out.

A heading is a small tunnel, and if timbered at all, it is timbered quite differently from the tunnel itself.

A heading 7 feet square is quite large enough for working skips or small trolleys.   If it is decided to get ordinary tip-waggons into the works, the heading must be at least 9 feet square.

If the ground is in rock, only occasional timbers may be required to support the roof when stones in the top are loose, or sound, when struck, what the miners call ' drummy.'   The top in that case may require 'head-trees,' which are timbers across the line of the heading, and polling-boards.

The head-trees may be let into the rock itself, or

may be supported on short props resting on ledges of the rock, which, from the shape in which they are put in, are called 'sprag-props,' *i.e.*, short spreading props. Or the head-trees may require to be supported by 'side-trees' the whole height of the heading; and these side-trees may, if the ground is soft in the bottom, require 'footblocks,' or 'sills,' which latter are timbers laid across the bottom of the heading from side to side.

In driving in clay or chalk, or in the shales or marls at the Severn Tunnel, it is possible to drive a kind of benching in the top, 1 foot in advance of the face, into which the head-trees are rolled and wedged in. If boards are required to support the roof, they are put in at the same time. This head-tree rests upon the ground of the face itself; then if side-trees are required two chambers are cut, one on each side into which the side-trees are inserted; the weight in the meantime resting on the ground of the face, which has not been disturbed, or a temporary prop in the middle. When the side-trees are in place, the head-trees are wedged up from them, and the small wedges which are used in this and other places in timbering the tunnel are called 'jacky pages' (perhaps 'jack up edges').

By these means the heading can be advanced in ordinary ground at a speed varying from 12 to 20 yards per week, allowing three or four months from the commencement to get the heading well advanced, and start what are called 'break-ups';

— HEADINGS —

— HEADING WITH HEAD TREES ONLY —   — HEADING WITH SPRAG PROPS —   — HEADING WITH SIDE TREES —

— HEADING WITH SILLS —   — TIMBERING IN COAL SHALE —

— *Curb for 29 feet diar Shaft* —

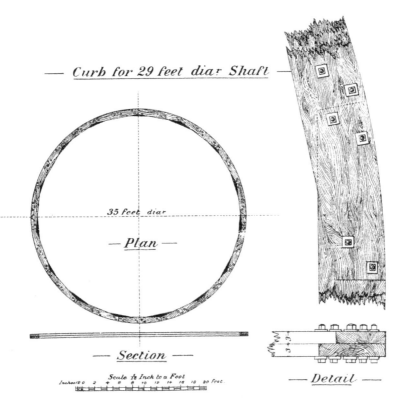

35 feet diar

— *Plan* —

— *Section* —

Scale ⅛ Inch to a Foot
Inches 12 0   2   4   8   8   10   12   14   16   18   20 feet

— *Detail* —

Machure & Cᵒ Lithⁿᵉ London.   LONDON   RICHARD BENTLEY & SON, 1887.

the tunnel itself can be completed at the same speed, Progress of the work.
the number of break-ups and the distances between
them being regulated with a view to this. 1881.

A 'break-up' is a hole driven in a slanting direction
from the top of the heading to the top of the tunnel
itself, or, if the ground is bad, to a height of
18 inches or 2 feet above that. When the break-up
is completed to the top of the tunnel, a top-heading
is commenced similar to the bottom heading, but
smaller in dimensions.

It may be taken to be about 3 or 4 feet wide at
the top and 5 to 6 feet wide at the bottom, and 6 feet
high.

The nature of the ground will determine the length
to be taken out at once ; whatever is decided upon
is called a 'length.' In very bad ground this may
be only 6 or 9 feet ; but generally it will not be less
than 12 feet, and, if in very strong ground, 20 feet
or even 24 feet.

When the top heading is completed to the required length, a benching is cut in the top of the
heading at one side, sufficiently large to receive the
timbers that have to carry the roof. These timbers
should be of good, fresh, round larch, varying from
12 inches to 16 inches in diameter, and are known
among miners as 'bars'; and the bars of the length
are divided into 'crown-bars' and 'side-bars.'

When one bar has been rolled on to the benching
prepared for it, another benching is cut upon the
other side of the heading, and a second bar placed

upon that ; between these bars the roof is secured by polling-boards, the thickness of the boards varying from 1 to 3 inches, according to the weight of the ground to be carried.

When the two bars are placed as above, chambers are cut down to the floor of the heading at a slight angle each way from the length, and into these chambers are inserted props, which are tightly wedged to the bars ; these props are known as ' back-props.'

Having secured the two first bars in the manner mentioned, another chamber is cut following the section of the tunnel like a shelf on the one side of the heading, into which the third crown-bar is rolled, and then secured by back-props in the same manner as the first, the fourth, and other bars following ; and when sufficient crown-bars are in this way inserted, secured by the back-props, and themselves holding the polling boards in place, they are secured by stretchers or short timbers ; between these a cutting is made, just inside the back-props, to the level fixed upon for the first sill. The sill-timbers are then brought in, in two pieces. They are large square timbers, from 12 to 15 inches square, scarfed in the middle, and when placed upon the ground, in the space which has been taken out for them, are jointed together with strong iron straps, called ' glands.' Another row of props is then inserted under the crown-bars, one under each bar resting on the sill ; and the same operation is continued till

BREAKING UP FROM HEADING IN SOFT GROUND.

CROWN BAR FIXED.

TOP CILL IN POSITION.

SCALE

INCHES 12 0   2   4   6   8   10   12   14   16   18   20   22   24   26   28   30 FEET

Maclure & Co Lith&c London.

LONDON. RICHARD BENTLEY & SON, 1887.

the whole tunnel is completed to the top of the

invert.

Two sills may not always be necessary, but the above description deals with ground for two sills; the working down fron₁ the upper sill to the lower one being exactly similar to that required for working to the first sill.

If the ground is good the crown-bars are placed entirely inside the brickwork of the tunnel, and they should not be so large as to equal half the space to be occupied by the brickwork. If the ground requires larger timbers than half the thickness of the brickwork, the bars must be worked as what are called 'drawing bars,' that is, bars to be drawn on end as the brickwork progresses; or must be placed entirely above the brickwork of the tunnel and built in.

When the length is completed, and the invert taken out, two profiles of boards— that is, light frames representing the exact shape of the invert—are set at the right level and to the right lines by the engineers; the bricklayers then commence and build the invert of the tunnel. When that is completed, what are called side-frames are set to guide the bricklayers in laying the bricks for the side-walls of the tunnel up to the springing of the arch. When the side-walls of the tunnel are completed to springing, the 'centres' are set.

The centres adopted at the Severn Tunnel were wholly what are known as 'skeleton centres;' that is, centres made by bolting two or three thicknesses of elm planks together into an arch of 3 inches less radius than the interior of the brickwork of the tunnel. The centres would be set from 3 feet 6 inches to 4 feet apart, according to the weight they would have to carry ; and the end ribs of each length should generally be thicker and stronger than the middle ribs, as they have to bear the weight of the crown-bars of the adjoining length. These end ribs are called 'leading-ribs.'

When the centres have been set, one 'lagging' is placed on at each side. A lagging is an ordinary plank or batten, 6 or 7 inches wide, and 3 inches thick. Behind this the bricklayers lay over-handed the bricks of the arch of the tunnel, making good to the ground at the back of the arch as they come up ; or if the ground is loose, to the polling-boards, which have to be left in. This system is continued on both sides at once, putting on one lagging at a time till the arch is completed, except about 18 inches in the crown. To complete this, which can only be done by one man, cross laggings are used, called 'block laggings.'

These being very short, are only made of $1\frac{1}{2}$-inch boards, resting upon the top laggings at each side ; the bricklayer, placing one or two at a time, works himself backwards till at last he completes the length, or, at a junction between two lengths, comes into a small hole, just the size of his body, which is known as the 'pigeon-hole.' This pigeon-hole he

— BREAK UP LENGTH WITH CENTRES —

TOP
HEADING

BOTTOM
HEADING

BARS TO BE DRAWN

Slack Blocks

Salt timber

8 Long bar

28. 0' dia

TOP
HEADING

BOTTOM
HEADING

SCALE

Inches 12 6 0 1 2 3 4 5 6 7 8 9 10 11 12 13 14 15 16 Feet

LONDON. RICHARD BENTLEY & SON, 1867.

Maclure & Co Lith. London.

has to fill up, and complete the arch brick by brick, working himself out of it.

When the arch is entirely bricked in a 'break-up length,' what is known as a 'running length' is commenced at one end of it. For this the top heading has been already driven, as before described, and then the bars are placed in the same manner as in the break-up length, propped with back props, and then with props upon sills; but, as one end of the bars rests upon the brickwork of the finished length, there is only one face to timber down. The face of each break-up and running length requires to be carefully timbered, and if the ground is loose, polled; and the sills require long struts to secure them against any pressure at the face itself.

When the running length is completed at one end of the break-up, another running length should be taken out at the other end, the miners working in one length, while the bricklayers are working in the other.

The advantage of the bottom heading is this: that all the mining done in every length is dropped down into the skips or waggons with but little expense in filling; and as the bottom heading can be pushed forward at the rate of from 12 to 20 yards per week, about every six or seven weeks fresh break-ups can be started at distances of 100 yards apart.

Of course the distance between break-ups must be regulated by the speed it is desired to attain in completing the tunnel.

If the ground at the top of the heading or tunnel

is loose, especially if it is full of water, a different system must be followed to secure the top.

Before mining to place a head-tree, or a crown-bar, polling-boards must be driven by mauls as piles; and where the ground is wet, this is one of the most difficult operations miners have to perform, for the dirty water streams over the end of the piles, and at every blow of the maul is spattered on all the men that are near.

A considerable length of the Severn Tunnel, on the Gloucestershire side, was in loose gravel, full of water, and required this operation.

In the same ground the crown-bars had to be placed entirely outside the tunnel, and the brickwork of the arch completed under them, the space between the crown-bars being filled up to the polling-boards with rough brickwork or rubble.

In shaft-sinking, I have known cases, more than thirty years ago, where, when a depth, say 20 or 30 feet, had been sunk from the top, a curb was placed, carried by iron rods, and in some cases by chains from timbers laid across the top of the shaft, and the brickwork for lining the shaft was built upon this curb ; and in some cases the brickwork has been built continuously on the top of the shaft, and the lining lowered away till it reached the required level. In other cases short lengths have been taken out in the sinking, and the brickwork added below the first curb on other curbs placed from time to time.

TIMBERING IN SOFT GROUND

BALLAST.

MARL.

BOTTOM
HEADING

HARP FOR BORING MACHINE

SCALE

Inches 18 0 2 4 6 8 10 12 14 16 18 20 22 24

SIDE VIEW

END VIEW

PLAN

SCALE

Inches 12   1   2   3   4   5 feet

LONDON. RICHARD BENTLEY & SON, 1887.

Maclure & C? Lith? London.

I have, however, adopted the system of completing the whole of the sinking of the shaft before
commencing any of the brickwork. In doing this,
strong curbs are made similar to the centres of the
tunnel; those for an 18-ft. shaft being 21 feet in
external diameter; those for a 15-ft. shaft being
18 feet in external diameter; and as we ultimately
sunk a large pumping-shaft, 29 feet in diameter,
with 3 feet of brickwork, the curbs in that case were
35 feet in external diameter.

These curbs are placed from 3 to 5 feet apart as
the sinking of the shaft progresses, with polling-
boards behind them to support the ground; and
when the shaft is completed to the bottom the brickwork is commenced, and each curb taken out as the
brickwork built up from the bottom reaches it.

In sinking a shaft at any place on the Severn Tunnel, the principal expense was incurred in keeping the
shaft free from water. We have tried all kinds of
pumps for this purpose, and there are objections and
difficulties with all.

The pulsometer-pump, which can be slung upon
chains and lowered as the work progresses, with a
flexible rubber-hose at the bottom of the pump, can
only be used to a depth of about 50 feet, and even
at that depth consumes so much steam as to be very
expensive.

A direct-acting steam-pump, of which we had
many patterns, made by a variety of makers, could
also be lowered in the shaft as the work progressed.

It is continually getting out of repair, on account of the quantity of grit and dirt to be raised with the water.

A chain-pump can be used only for low lifts, say of 30 or 40 feet. Even the ordinary bucket-pump, which is the only one that can be trusted for working to depths of 100 to 200 feet, is subject to be constantly damaged and broken by the blasting in the pit. It is necessary, also, to have the holes of the suction-piece or wind-bore of the bucket-pump close to the bottom of the pump, or the miners would be standing in a considerable depth of water, and the pumps rapidly wear out their buckets from the quantity of sand and grit raised with the water.

In four of the shafts we sank at the Severn Tunnel, after we had a first shaft down and pumps fixed in it, we escaped this pumping difficulty by driving a heading from the existing pumping-shaft to the shaft we were sinking, or rather to the spot where the shaft would eventually be when it should be sunk its full depth. From this heading we sunk the bottom part of the shaft and lined it with brick-work, and then put down a bore-hole through the upper strata till it reached the completed work below; the water then drained through this bore-hole, as the men sunk the shaft, into the heading beneath, and so into the other shaft alongside from which it was pumped. In two of the shafts, by driving this cross-heading before the bore-hole was made, and accurately setting out the centre of the shaft

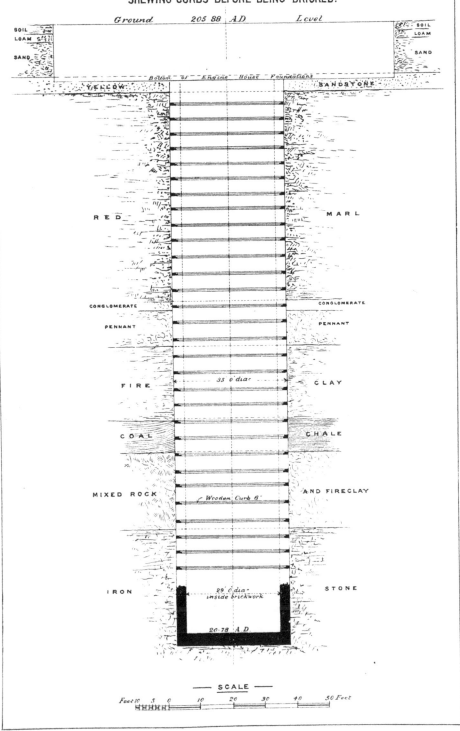

PERMANENT PUMPING SHAFT AT SUDBROOK
SHEWING CURBS BEFORE BEING BRICKED.

Ground   205 88 AD   *Level*

SOIL
LOAM
SAND

SOIL
LOAM
SAND

Bottom of Engine House Foundations

YELLOW   SANDSTONE

RED   MARL

CONGLOMERATE   CONGLOMERATE

PENNANT   PENNANT

FIRE   *35 0 diar*   CLAY

COAL   SHALE

MIXED ROCK   *Wooden Curb 6"*   AND FIRECLAY

IRON   *29 0 dia inside brickwork*   STONE

20 78 A D

SCALE

Feet 10   5   0   10   20   30   40   50 Feet

LONDON   RICHARD BENTLEY & SON, 1887

underground through this heading, we succeeded in

putting in the bottom and 20 or 30 feet of the shafts without having any water to contend with.

In a heading being driven at any distance from the shaft, though there may be but five or six men in the heading, the heat becomes very great, and, of course, a constant supply of fresh air is required for the men themselves.

To supply this, air-compressing engines were worked upon the surface. Besides some smaller ones, which had been provided by the Great Western Railway Company before the contract was let to me, we had two double 20-inch air-compressors, and one double 16-inch air-compressor. One double 20-inch was worked at the main shaft at Sudbrook, one 20-inch at 5 miles 4 chains, and the 16-inch one at the Marsh Pit.

The air from the compressors was led to an air-receiver, a wrought-iron cylinder, something like a boiler, in which the air was maintained at a pressure of about 60 lbs. to the square inch. Under this pressure the air was so hot that it was hardly possible to bear the hand for any time upon the outside of the receiver. From the receiver wrought-iron steam-pipes, 2 or 3 inches in diameter, were led away to the various places where rock-drills or other machinery were to work.

As almost all the headings required blasting, the boring of the holes was done by machine-drills. Several patterns of these were used: principally drills made by the Great Western Railway Company at Swindon, from a pattern of their own. This proved to be an excellent drill for making rapid progress ; but it was complicated in design, and required constant repairs.

Another drill used was the 'Darlington,' which we liked better, because it required so little repairs. One of these drills could work for six months, after being started, without being repaired at all.

In addition to these we had several others, among them the 'Duncan' drill, which did excellent work.

Several methods have been tried for holding the drills in working a heading.

I think decidedly the best is the system adopted by the Great Western Company originally, of a heavy casting, called, from its shape, a 'harp,' with sliding bars that can be placed at any angle, and each bar carrying drills. These harps were made to run upon a strong frame on flange wheels of the same gauge as the road used for the skip trolleys.

They could be pushed forward into the face, and when the holes were drilled, and before they were charged and fired, could be run back a short distance, out of danger from the blasting, and brought forward again when the material brought down by the blasting had been cleared up.

We used all kinds of explosives in the work : dynamite, gelatine, tonite, gun-cotton, compressed powder, and ordinary powder.

Powder made so much smoke as to be very objectionable in the workings, and was not strong enough

for the hard rock.

The fumes of dynamite were so deleterious, and even dangerous, that we abandoned the use of it altogether.

Gelatine produced the best results in the hard rock, but we only used it for some months, when the makers were required by the Board of Trade to suspend the sale of it, and return all the material to store.

Nearly all the work was done by tonite made by the Cotton Powder Company. This is a carefully prepared explosive, made up into handy packages, which the miners know as 'pills,' with a detonator attached to some of the cartridges, which were then known as 'primers.'

The fuse was attached to a primer, and then, according to the depth of the hole and the strength of the material to be blasted, a primer and one, two, three, or four pills were placed in the hole.

Very little tamping was required ; but in very hard rock the tonite would leave what the miners call a 'socket ;' that is, a short length of the hole at the bottom where the explosive had not been strong enough to dislodge the rock. When using gelatine we found no sockets left, showing the extra strength of that explosive.

The tonite was not injured by water, nor did it seem to be affected by cold, as dynamite is ; and the fumes were so slight that it was quite possible to return to the face within a minute from the time the shots were fired.

When the works were about half completed, I was persuaded to try compressed lime-cartridges, with which it was stated good work was being done in the coal-mines.

A number of these cartridges were brought down and tried in various places, but with no favourable result.

Generally, in ground so hard as that we had to deal with, the lime only blew out its own tamping, and displaced no rock   The experiment, too, was attended with an unfortunate accident   The cartridges of compressed lime are placed in a large bore-hole, drilled to receive them, and a tube, instead of a fuse, is placed in the hole above the cartridges, and when the tamping has been placed, water is forced by a force-pump through the tube on to the cartridges.   The swelling of the lime and the generation of gas is to displace the rock.

When one of the holes had just been loaded, my principal foreman, Joseph Talbot, was standing opposite to the hole, and the first of the water had just been pumped in, when the tamping blew out and a quantity of lime was forced into his eyes, and for a day or two it was feared he would lose his sight.   Fortunately he entirely recovered ; but the short experience was quite enough to convince us that we could make no use of the lime-cartridge system.

When the machine-drills were being worked by

compressed air in any of the headings, the air, after passing through the drills, kept up efficient ventilation.

We afterwards applied this compressed air to the pumping. When we were sinking the drainage-heading, 20 feet below the old workings, it was done from a number of small shafts sunk to a depth of about 25 feet from the old heading ; and to pump the water from the lower level I adopted a pump which was designed by Mr. A. O. Schenk, and which proved to be thoroughly efficient. The pump consisted of a wrought-iron closed tank of any suitable dimensions, say about 3 feet square. A 6-inch wrought-iron pipe, passing from near the bottom through the top of the tank, was led up to the level at which the water was to be delivered. A $1\frac{1}{2}$-inch wrought-iron pipe was connected from the compressed air-main to the top of the tank. At the side of the tank, near the bottom, several flap-valves were fixed, opening inwards.

The tank was then lowered into the bottom of the shaft, which was 4 or 5 feet below the level of the bottom heading, and the connection with the air-pipe was made. The water flowed into the tank through the flap-valves till the tank was full; the pressure of the air being then admitted at the top of the tank, it closed the flap-valves and forced the water up the 6-inch pipe to the required level ; when the tank was empty the air-pressure was turned off, and the tank filled itself again.

The only thing necessary was to make this self-acting, and this was done by Mr. Schenk in a very ingenious manner. A small wooden box, divided into two compartments, was fixed to rock on a centre at the top of the shaft, and a small pipe led to discharge directly over its centre. When this small pipe filled the one compartment of the box, so that it was heavier than the other, the box turned upon its centre-pins, and in doing so opened a valve on the air-main by means of rods attached to it, and allowed the compressed air to pass into the tank and force the water out of it to the surface. The small pipe then filled the other compartment of the box, and reversed the action, so that the pump was made perfectly self-acting, and required no looking after. These pumps were used throughout the period we were lowering the heading, and proved a complete success; and afterwards, when we had rather more water at the Marsh Pit than three 15-inch pumps could master, I fixed one of them there to lift water to a height of 100 feet. It was as perfect a success lifting water 100 feet as it had been lifting 25 feet.

Of course, working by compressed air in this manner is expensive; but underground, on account of the difficulty of ventilation and the heat, it is necessary in many places to adopt compressed air for working machinery. Electricity could be used for the same purposes, but I have only used the compressed air, and that I have used for winding by steam-crabs, for pumping, and working rock-drills.

When the heading is driven, and a number of

break-ups commenced along it, the great difficulty will be to take out the requisite number of skips of rock or other material filled from the lengths and the heading-face, and to take in the empty skips to be loaded, and to carry to the bricklayers the bricks and cement which they require ; while at the same time there will constantly be timbers to be taken in to the miners.

Where the length to be travelled was short, this presented no great difficulty ; but under the river about a mile of the work was done eastward from the shaft at Sudbrook, and about $1\frac{1}{4}$ miles westward from the shaft at Sea-Wall.

For the length done from Sudbrook, the gradient falling towards the shaft, there was laid up the 9 ft. barrel, throughout the whole of its length, a double road of 1-ft. 9-in. gauge, so that the skips could travel on the up or down road ; but beyond the end of the 9-ft. barrel, where the break-ups were commenced, the road had to be arranged in a series of single roads and turn-outs or sidings, to allow the skips to pass each other. The roads being thus arranged, the hauling was done by stout ponies from 13 to $13\frac{1}{2}$ hands high, and these ponies became most intelligent at their work, and knew exactly what to do, even as well as the men themselves.

On the other side of the river, however, where all the material had to be taken out up a gradient of 1 in 100, it was much more difficult to take out the excavated material and to keep up the supply of bricks, cement, and timber; and as soon as the first length of half a mile of tunnel was completed, I put down a hauling-engine at the top of the Sea-Wall Shaft to take out the skips and bring in the bricks and cement. This hauling-engine had two cylinders, each 12 inches in diameter, and worked a large pulley, round which three turns of a wire-rope or bond were taken; the two ends after leaving the gulley descended the shaft, and there passed round other pulleys to change the direction. They then ran down the tunnel, first for half a mile, and eventually for a distance of rather more than a mile from the shaft, and at the extreme end passed round another pulley, and the roads on which the skips travelled were laid as far apart as the diameter of this large pulley. Steel rollers were fixed on the sleepers of these roads several feet apart, over which the 'bond' ran; and at the top of the shaft there was attached to the rope a heavy trolley, carrying a considerable weight of iron placed upon a sharp incline; the weight of this trolley on the incline serving always to keep the wire-rope sufficiently tight. When the engine was started there was an endless wire-rope constantly in motion at a speed of about 2 miles an hour; the one line of wire-rope running down the tunnel and the other line up. The skips were brought out by ponies or men to the end of the rope, and there attached by the 'hookers-on,' by means of a clip, to the wire-rope, without stopping it.

The descending skips were pushed off the cage at

the bottom of the pit, and were in like manner
attached to the descending rope by another man.
The skips were thus attached to the rope at any
irregular intervals just as they came, without loss of
time ; and sometimes there were as many as 100
skips on the ascending rope and 100 on the descending one.

We found the system to work perfectly, and it reduced the cost of hauling out to less than half the
cost of ponies.

CHAPTER VII.

A YEAR OF GOOD PROGRESS [1882].

On the 2nd November, 1881, I received orders from Sir John Hawkshaw to go on with the whole of the works as rapidly as possible. The contract had contemplated that for the first eighteen months only the work under the 'Shoots' should be proceeded with ; but it took as nearly as possible twelve months from the date of signing the contract to clear the works of water. Twelve months more had been used in completing the heading under the river, securing the old shafts, sinking new ones, and commencing the brickwork under the 'Shoots' and under the Salmon Pool ; but sufficient had been done to inspire confidence, and the order was therefore given to proceed with the whole of the works at once. I therefore purchased, at the end of December, an additional $4\frac{1}{2}$ acres of land for various purposes, and leased a further quantity of $3\frac{1}{2}$ acres for building additional houses.

At the end of the year 1881 about 60 yards of the full-sized tunnel had been completed at the bottom of the Sea-Wall Shaft, and about 300 yards of con-

tinuous arching had been turned westwards of the shaft.

In addition to this the arching had been commenced at two break-ups going westward, the long heading had been joined, and ventilation was established through the works between the Monmouthshire and Gloucestershire sides of the river. Four break-ups had been started in the Pennant rock under the 'Shoots,' and about 100 yards of arching had been completed from these four break-ups.

The 9-ft. barrel had been entirely lined with brickwork, and the 18-ft. tunnel at the bottom of the new shaft at Sudbrook had also been completed with a bellmouth into it, out of the 9 ft.

A head-wall had been built across the entrance to the Iron Pit, and a 3-ft. sluice-valve fixed in it, from which rods were led up the New Pit to the surface of the ground.

About 60 yards of tunnel had been completed at the bottom of the New Pit, and both the New Pit and the Old Pit at Sudbrook had been secured by brickwork to the arch of the tunnel.

At 5 miles 4 chains both shafts had been bricked, and about 15 yards of tunnel completed at the bottom of the winding shaft.

The brickmaking plant had been established and started to work.

The number of houses provided for the men was one detached house for the principal foreman, one for the Company's chief inspector, one for the chief storekeeper; six large cottages, and six smaller ones adjoining the Roman Camp; six semi-detached houses for foremen; a large coffee-house with a reading-room, and an adjoining house for the man keeping the coffee-room; twenty stone cottages closely adjoining the main shaft; and two stone semi-detached houses for foremen.

A mission-room with 250 sittings had also been completed, with schoolrooms behind, in which a day-school was carried on throughout the whole of the year; and at the end of the year a separate detached schoolroom of three rooms had been built, a certificated master appointed, and the school was put under Government inspection.

The principal office for the works, a large two-storied building, had been built, with a cottage adjoining, for the residence of the office keeper.

At each shaft large rooms called 'cabins' had been built for the men to take their meals in, and where they could dry their clothes.

A saw-mill had been established, and a large carpenter's shop.

Stables for twenty horses, with a cottage adjoining, had been built; also a new fitting-shop and blacksmith's shops.

The roads to connect the new houses with the main roads had been completed, and in addition to the permanent houses, six wooden houses had been built near the brickyard for the men employed there.

On the Gloucestershire side of the river, where

there was a difficulty in obtaining any land except directly over the tunnel, a number of wooden houses had been built, following the line of the tunnel. At this point I was afraid to put up brick houses, because the tunnel was to be constructed in gravel, and there would probably be serious settlement when the work was being carried on.

As soon as the 9-ft. barrel was completely bricked, in the middle of this year, a number of shafts were sunk to a depth of 20 feet, from which to put in the permanent drain required by the alteration of gradient. This when finished was to be a 5-ft. barrel.

The ground in which the drain was constructed was principally hard Pennant rock, but for a short distance it passed through the coal-shale. No great quantity of water was met with, and for the first 1,000 feet this was easily disposed of by hand-pumps. From the end of the 1,000 feet the water from this drain was pumped by the Schenk-pumps, which have been mentioned before.

The permission to commence the works at all points enabled us early in 1882 to largely increase the number of men employed; and as soon as the land was purchased, we commenced the open cuttings on both sides of the river.

The progress made during the whole of this year was very good, and there is little to record with regard to it, except that now and then some of those accidents, which are unavoidable in tunnel-works, occurred.

The first of these took place about 60 yards east from the New Pit at Sudbrook, when we were getting out a length of tunnel in the last week of January, 1882.

The upper part of this length was entirely in coal-shale. The length was nearly ready for the brick-layers, and looked all secure, when a large mass of coal-shale in the face slipped out off a concealed bed of rock, which stood at an angle of about 45 degrees, and in slipping knocked out the whole of the props under the sills. The knocking out of these props caused the sills to break; the upper part of the face then also slipped in, and we 'lost the length.' This was the only length we lost in the whole tunnel; and as the total number of lengths taken out was over 1,500, it shows that great care was exercised by the foremen and the miners, to be able to say that only one length of the 1,500 was lost.

The losing of a length in such ground was, how-ever, a serious matter.

It was under the river, but fortunately there were the hard beds of Pennant and conglomerate above the coal-shale, so that we had no reason to fear that the water would break into us.

As quickly as possible the top was secured, and then one of the most difficult operations in mining was commenced, viz., to pole through the broken ground.

SIR JOHN HAWKSHAW INSPECTING THE WORKS.

Machre & Co. Lith.rs London.

AT WORK IN THE TUNNEL.

LONDON   RICHARD BENTLEY & SON, 1887

Whereas the length itself had been taken out in a fortnight, it took more than two months to take it out the second time.

Towards the end of February, Sir John Hawkshaw for the first time passed through the whole of the heading under the river. The work was then to be seen in all stages of progress. The old timbering, which had been put in before 1879, was also to be seen; and the points where the roof had fallen in, in consequence of insufficient timbering.

There were at that time about ten break-ups at work, half of them in hard rock, and half of them, on the east side of the river, in the red marl.

Early in April the brush electric light system was brought into operation on the works on the Gloucestershire side of the river. This was a 12-light system, and was used for lighting the top and the bottom of the pit, the spoil-bank on which the skips were tipped, and the length of tunnel where the arch was completed; each light was of 2,000 candle-power.

Shortly afterwards, a 40-light system was started on the Monmouthshire side of the river. The cable from this dynamo was 5 miles long, lighting the yard at Sudbrook, and the top of the pits there, the roads leading to the works, the top and the bottom of the pits at 5 miles 4 chains, and the top and bottom of the pits at both Marsh and Hill; and wherever the arch of the tunnel was completed for

a distance of 100 yards, the lights were also insti-
tuted underground.

Electric lighting was comparatively in its infancy
at the time this plant was erected, and I was advised
that it was preferable to lay the cable (which was an
insulated cable containing seven copper wires) in
boxes underground, buried in cement concrete.   It
was supposed that there would be considerable danger
to the men if the cables, though insulated, were carried
upon poles above the surface of the ground; but
a very short experience convinced us that it was
impossible to work with the cables underground.

When any quantity of water made its way into
the boxes, a current was set up between the two
cables, and the wires rapidly fused.   In consequence
of this, a great deal of trouble was experienced with
the light at first; and ultimately the whole of the
cable was taken up, and fixed upon poles at least
15 feet above the surface of the ground, when the
light was found to act admirably.

This light was a great advantage both for working
on the surface at night and for the men running out
skips from the lengths below—one great advantage
being that it generated little or no heat; but of course
we could not use these lamps in the lengths that were
at work on account of the blasting.   And even in
those places where a man could work by the electric
light, we generally found that he placed a candle in
front of him on account of the deep shadow thrown
by the light from his own body.

In the course of this year we also established a

telephone from the works under the river to the Gloucestershire side. The first day it was at work it was, I believe, the means of averting a strike, for just as the principal foreman happened to go into one cabin, a discontented ganger entered the other, swearing and grumbling, and saying what he would do if some fancied grievance were not put right, and what he would advise the men to do. He little thought all he was saying was heard by his ' boss '; but his instant dismissal prevented further mischief.

During the year 1880 the progress which could be made with the works was only at the rate of about £4,000 worth of work done per month.

During the year 1881 the progress after the water was got out, but while the operations were confined to the work under the ' Shoots ' and to the new shafts, was only at the rate of about £7,000 per month ; but directly permission to proceed with the whole of the works was given, the progress rose rapidly, and was £11,000 in the month of January, 1882, and £23,000 per month before the end of the year.

In the beginning of 1882 one of the largest of the cottages had been temporarily converted into a cottage hospital, with a nurse employed to carry on a proper system of nursing under the doctor in charge.

An arrangement had been made with Dr. Lawrence, one of the leading doctors in Chepstow, to take entire charge of the men upon the work, and to keep a resident assistant there. Early in the summer the erection of a separate cottage hospital had been commenced, which included a residence for a matron, rooms for the resident doctor, for a sister to superintend the nursing, and for an assistant nurse.

This hospital was completed and opened in the second week in October. A plan showing the arrangement of the wards and the dwelling-house is given.

Considering the magnitude of the undertaking, the difficulties encountered, and the number of men employed night and day, we were very free from accidents during the six years the works were in progress ; but still we found the hospital of the greatest value in treating both accidents and diseases, such as congestion of the lungs, rheumatic fever, etc. The principal illness that the men suffered from was pneumonia, caused no doubt by the great heat and damp below, and then careless exposure when they came out of the works.

Besides the general wards in the hospital, we had an operating-room, an emergency-ward, and a ward for women and children.

It has been before stated that the mission-room to hold 250 was opened in the end of the year 1880. By the end of November, 1881, this room was so crowded that it became necessary to take down the partition which separated the schoolrooms from the mission-room, to remove the day-school entirely into

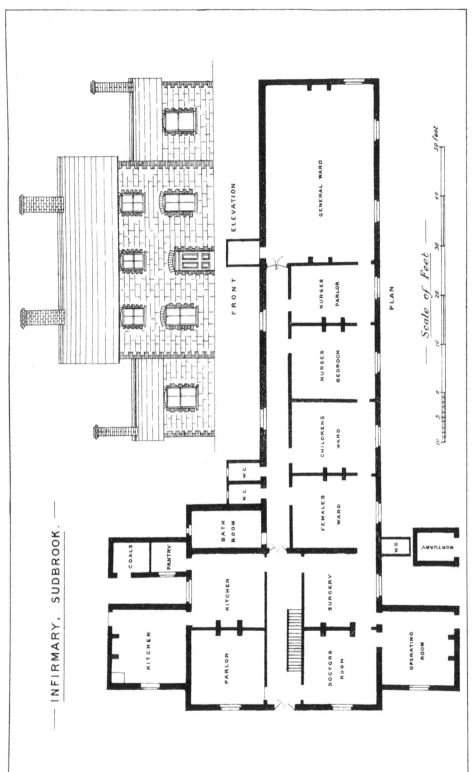

— INFIRMARY, SUDBROOK. —

KITCHEN

COALS

PANTRY

BATH ROOM

W.C. W.C.

PARLOR

KITCHEN

SURGERY

DOCTORS ROOM

OPERATING ROOM

W.C

MORTUARY

FEMALES WARD

CHILDRENS WARD

NURSES BEDROOM

NURSES PARLOR

GENERAL WARD

PLAN

FRONT ELEVATION.

— Scale of Feet —

10 5 0 10 20 30 40 50 feet

LONDON RICHARD BENTLEY & SON, 1887.

Maclure & Co Ltd. London

the new rooms that had been built for it, and to

enlarge the mission-room to hold 400. By the middle of November, 1882, this room was crowded, nearly 500 people attending every Sunday evening, the people sitting all up the aisles, and a number standing. I had a plan made for enlarging this room, but the difficulties in the way of doing so were very great, because it stood in close proximity to the new day-school, and any enlargement would take away the light from the windows of the school. The day-school itself was also crowded, and I was in great doubt as to what was the proper thing to do under the circumstances. The mission-room was heated by hot-air flues passing under the aisles ; and after service on the 26th November, 1882, when the room had been terribly over-crowded and exceedingly hot, a fire broke out in the night, and the whole building was burned to the ground.

A policeman who passed on his rounds at midnight stated that he saw no signs of fire ; but about three o'clock in the morning the driver of one of the Cornish pumping-engines, looking out of the window of his engine-house, saw the mission-room alight from end to end. The alarm was immediately given, but the fire had already a complete hold of the room. All the seats and the roof were of pitch-pine, and in less than an hour the roof fell in and the room was destroyed. All the books and the American organ were destroyed with it.

I had already leased three acres of land, closely

adjoining this room, for building additional houses for the men; and on Monday morning, using to some extent the plans which had been prepared for enlarging the old room, we began a new and larger one. At ten o'clock the men commenced to dig the foundations. At one o'clock the masons started to build the walls. The electric light was put up to enable them to work night and day, and though frost interfered to some extent with the rapid progress of the work, a new mission-hall to seat 1,000 persons was completed on the 16th December, and opened for service on the 17th—less than three weeks from the time of the fire, so that the men were only kept two Sundays out of the room, and on these service was held in the large reading-room over the coffee-room, which was capable of holding 250 persons. A drawing is given of the new room.

Early in the year, the head-wall to the east of the Sea-Wall Shaft was removed, and the heading was commenced on the lowered gradient, going eastward from this shaft.

In order to work down to the lower gradient westward from this shaft, small shafts were sunk, at intervals of about 60 yards, from the old heading to the new levels, and a small heading, 6 feet by 4 feet, was driven from shaft to shaft, till we by this means secured natural drainage for putting in the invert from Sea-Wall Shaft, westwards.

Break-ups were started through the whole length of the ' Shoots,' and the ground proving very good,

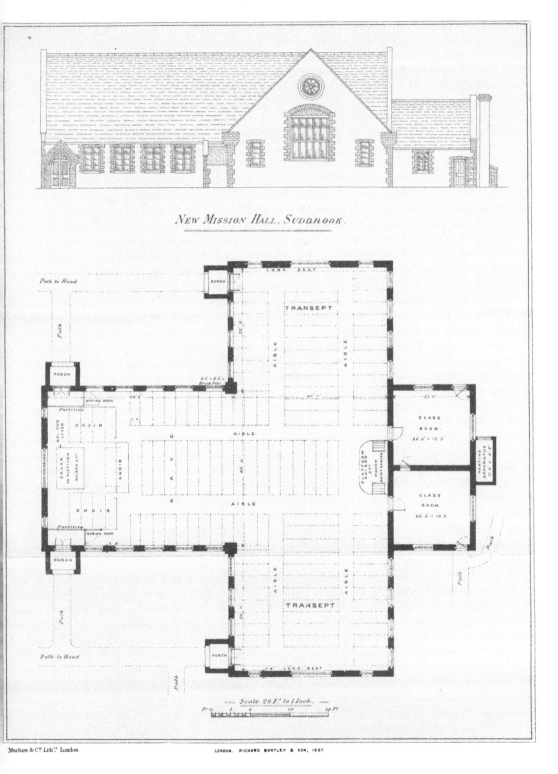

NEW MISSION HALL, SUDBROOK.

LONDON. RICHARD BENTLEY & SON, 1887.

The material originally positioned here is too large for reproduction in this reissue. A PDF can be downloaded from the web address given on page iv of this book, by clicking on 'Resources Available'.

lengths averaging 20 feet were taken out in the
Pennant rock, and the arch pushed on rapidly.
Although the rock was good at this point, and rapid
progress was made, no length was allowed to be
executed without timbering.

The arch was completed from 5 miles 4 chains
pit to within 30 feet of the point where the Great
Spring had broken in, in 1879. The heading was
pushed forward westward from 5 miles 4 chains,
and a break-up commenced in the conglomerate
rock.

Every yard of ground that was opened at this pit
gave additional quantities of water ; and I found
it necessary, in addition to the two 18-inch pumps
which were working there, to put down two 28-inch
plunger-pumps, worked by a large horizontal engine.

The Marsh Pit was lowered to the new gradient,
and the two 15-inch plunger-pumps were refixed there.
We had recommenced to drive the headings east and
west, and a length of about 60 yards of tunnel was
completed adjoining the pit. The old heading was
lowered to the new gradient for a distance of about
200 yards westward from the pit, and a break-up had
been commenced about 100 yards west of the pit.

The Hill Shaft had also been lowered to the
altered levels, and two 15-inch plunger-pumps had
been fixed in that shaft, with a new pumping-engine ;
and the heading had also been lowered for a distance
of about 150 yards west of the pit, and a break-up
started about 100 yards from the pit.

Progress of
the work.
————
1882.

In the month of May, the large open cutting on the Gloucestershire side of the river was commenced.

In the original contract the quantity to be excavated in this cutting was only 366,000 yards; but by the lowering of the gradient the quantity had been increased to upwards of 800,000 yards. About 200,000 yards of this was to be used in forming the sea bank around the cutting to protect it from any extraordinary tides, and this work, which was done by wheeling out from the sides of the cutting, was carried on through the whole of the year.

THE PANIC.

On Saturday, the 2nd December, six days after the mission-room had been destroyed by fire, I had been to the hospital to see some men who were there, when, coming out of the hospital just before one o'clock, I was met by one of my people from the office, with a face exhibiting the most complete signs of terror. On asking what was the matter, he said: 'The river is in, the tunnel is in!' and this was all the answer I could get.

'Where are the men?'

'They are just coming up the shaft.'

I hurried to the top of the main shaft, and there I found between 300 and 400 men, evidently in the greatest terror and distress. Some had lost part of their clothing; hardly one of them could speak from exhaustion; and they were anxiously watching for the arrival of the large cage, which was bringing up

a further batch of men.

Every man was panting for breath, and excited to the last degree with fear.

I must say that my heart sank, and I feared the worst; but at that moment the cage arrived at the top with ten or twelve men, and a foreman, named Tommy Lester, who I knew had been working beyond the 'Shoots.' I turned to him eagerly, and said:

'Lester, what did you see?'

'I see nothin', sir.'

'What is it, then?'

'I don't know, only the river's in.'

'Where were you working?'

'In No. 8.'

'And you saw nothing?'

'No. It was beyond me.'

I turned to another, and said:

'Where were you working?'

'In the long heading.'

'And what did you see?'

'I see nothin', but the river's in.'

'I think you are a pack of fools,' I said; 'I think a five-pound note will cure it all.'

My reason for saying this was that these points were beyond the 'Shoots,' and there the bed of the river was dry at low water; so that if any hole existed there we could stop it as we had done the hole in the Salmon Pool.

Just then the foreman of the Cornish pumps came up, and said :

'I cannot understand it ; there is no more water at the pumps.'

I walked over with him and with Mr. Simpson, one of my staff, to the edge of the river, and found the water from the discharge-culvert the same colour as usual ; so I walked back to the pit, and said :

'I am going below ; who is going with me ?

J. H. Simpson and Jim Richards jumped into the cage with me, and the signal to lower was given. On arriving at the bottom we found it *perfectly dry*, and four or five men sitting quietly on a piece of timber scraping their boots.

The principal foreman of the works, Joseph Talbot, who had reached the top of the shaft before me, had gone down the iron staircase in the pumping-shaft, thinking that the winding-shaft could not be used, and he proceeded at once up the heading to see what was the matter.

After passing beyond the 9-ft. barrel where the men had been at work, in several break-ups he found articles of the men's clothing thrown in every direction—hats, neckerchiefs, leggings, waistcoats, everything which they could take off and throw away ; besides this, nothing was to be seen but the ponies which had been employed drawing the skips to the bottom of the shaft.

Returning again to the top, I explained the condition of affairs, and my readers may imagine that

there was no little chaff at the expense of the men who had run away under the influence of panic.

By degrees, by questioning one and another, the whole story was brought to light. The men had been working in break-ups, and also in extending the long heading on the new gradient. This heading passed under the old heading (being driven for some distance level, while the other rose 1 in 100) till it left sufficient ground overhead to commence a heading entirely under the original one. A con- siderable stream of water, about 2,000 gallons a minute, was always running down the old heading, and where the new heading started under the old one, the water was carried on one side in a wooden shoot. The upper part of this shoot was secured in a dam made of clay-puddle to prevent the water from falling over the face of the heading.

On the other side of the river, I have before stated that a length of bottom heading had been driven from a number of small shafts to reach the levels of the lowered gradient. The water flowing down this bottom heading rose up the last of the shafts, and a length of about 1,500 feet always had water in it. In making the junctions of this heading between the different shafts where there was con- siderable water, all that the men could do was to break down the last piece of rock between two shafts by blasting, without being able to go back to enlarge the hole made by the shots, or to clear up the rock which had been displaced by blasting ; the opening in the heading, therefore, between some of the shafts was probably very small, and some timber which had dropped down from the upper works had got into one of these holes and dammed back the water. The foreman on the Gloucestershire side, in order to let the water flow freely, had sent down a diver to remove this timber ; and when the timber was removed, the water which had been dammed back escaped rapidly, and flowing down the long heading had overflowed the puddle 'stank' made by the miners in the bottom heading. They had sent out twice to raise this stank, when they found the water rising above it, and at last one of them, seeing the water come still more rapidly, was seized with panic, and he called out :

'Escape for your lives, boys! the river's in !' and the men had taken the alarm at once.

As they ran towards the shaft, the men in the other break-ups joined in the panic, and at last the whole stream of men—three or four hundred in number—ran for their lives to the winding-shaft at Sudbrook.

When passing through lengths of finished tunnel, they spread out in a disorderly crowd, running perhaps 20 feet wide ; then they would come to a short length between two break-ups, where there was only a 7-ft. heading. Here they threw each other down, trampled upon each other, shouting and screaming ; and then, to add to the disorder, the

ponies in the various break-ups took the alarm,
and galloped down in the direction of the winding-
shaft, trampling on the prostrate bodies of the
men.

One ganger, a little stout fellow, who was working
at the 5-ft. barrel, was, on account of his short
legs and stout body, unable to keep up with the
others; and he was said to have uttered the most
piteous entreaties to the men to carry him, and not
leave him to be drowned.

The principal foreman afterwards asked one of
the men why he had thrown away his clothes.

'To zwim, zur,' he said.

I fear he would have made a bad hand at swim-
ming in a 7-ft. heading, if any large volume of water
from the river had really entered the works.

When the men reached the top of the pit, the
night-shift—which would go below at two o'clock—
had already received their pay, and were gathering
in preparation to descend. It may be imagined
that these men cruelly chaffed the others, who had
come up, as soon as it was known that there was no
danger below; and I have reason to think they
reaped quite a harvest of neckties and other things,
thrown away by the others, when they went down
to their work.

It would be wrong for anyone reading this account
to blame the men for cowardice. I am sure that a
finer body of men could not be found in England.
They were men who had had a thoroughly hard

training, and accustomed as they were to work in dangerous positions, and knowing every time they went below that each man took his life in his hand, they still went cheerfully to their work, and were no doubt as brave as Englishmen always are. But to understand how easily a panic spreads, under the circumstances, it would be necessary one's self to be under the river, a mile away from the shaft, confined in a narrow space with rocks dripping or running with water all round, with only the light of a stray candle here and there, and the most extraordinary sounds that ever greeted the ears of mortal man: first from the east, and then from the west, heavy timbers thrown down suddenly, with a noise that re-echoed through the whole of the works ; then a stray shot fired in one direction, then a complete salvo of 50 or 60 shots from the other—every sound totally different from the sounds in the open air—all the surroundings such as must produce a feeling of awe and tension of the nerves ; and then, when men following their dangerous employment heard others running by them below, shouting to them to escape for their lives for the river was in, would any man pause to consider, when he thought his life could only be saved by the rapidity of his flight from an enemy against which he could not contend ?

I do not blame the men for the panic ; but I had a bad quarter of an hour myself, when it seemed as though the three years' work was after all ending in failure.

We had also an amusing accident on the Gloucestershire side of the river just a fortnight before,
which no doubt caused the men more easily to take
alarm.

In driving the bottom heading from Sea-Wall
Shaft in the red marl, at a point where we had
reason to believe, from the borings which had been
taken, that there were still 6 or 7 feet of strong
marl over the roof of the heading, on the night of
the 13th November, the men being shorthanded
had stopped work at the face of the heading, and
had been drawn away to other parts, when (at midnight) the crust of marl at the top of the heading—
which proved to be only 6 inches thick, instead of
6 feet—had suddenly given way, and the gravel,
which was alive with water, had poured into the
bottom heading, and the wooden houses being
directly over the tunnel at this place, one large
brick chimney, which was between two houses, had
gone straight down, like a ramrod into the barrel of
a gun, into the heading below. The houses having
been built of wood, on account of this danger being
foreseen, were but slightly damaged, the floors being
slightly bent against the hole through which the
chimney had disappeared. The men, women, and
children occupying the two houses were asleep, and
none of them were injured; but it certainly was
ridiculous to hear of the number of trousers and
waistcoats, all containing money and watches, which
had been hung upon nails driven into this chimney

the night before, and which had gone down into the bowels of the earth.

By arrangement with the Post-Office authorities, we had in the course of this year obtained the establishment of a post-office, money-order office, and savings bank upon the works, and in the end of October we also secured a telegraph-office, which was a great accommodation and was largely used. The number of messages which passed through this office rose to 1,900 in 1883, 2,100 in 1884, and nearly 4,000 in 1886.

The Post-Office authorities, when this post-office was established, gave it the name of ' Sudbrook,' as they already had an office at Portskewett ; and the settlement henceforth took the name of Sudbrook, and assumed quite the appearance of a small town.

Fourteen stone houses were built this year in the neighbourhood of the brickyard. Twenty were built, between the brickyard and Sudbrook Shaft, of concrete ; and eighteen brick houses in further continuation of the same row. Four foremen's houses, and the post-office, and ten more wooden houses were erected.

A plan is given showing the progress made with these buildings, and the year 1882 closed with the opening of the new mission-hall.

THE WORKS. SUDBROOK.

LONDON. RICHARD BENTLEY & SON, 1887

RIVER SEVERN

PARISH OF PORTSKEWETT

Remains of Roman Camp

Site of Chapel found

to Bristol & Present

SUDBROOK

CAMP ROAD

POST OFFICE ROW

CHURCH ROW

STATION

800 R

120 CHAINS RADIUS

CENTRE LINE OF RAILWAY

THE VILLAS

Concrete Houses

SEA VIEW

5 MILES

SUDBROOK TERRACE

to South Wales

5. M. 4 C. BRICKYARD

DRYING SHED

REFERENCES

22	Carpenters' Shop.
23	Electric Light Shed.
24	No 1 70" Beam Engine House. 12 ft. [Shafts.]
25	Cement Shed.
26	Miner's Cabin.
27	School Room.
28	Brick Kilns.
29	Blacksmith's Shop.
30	Clay Mill.
31	Winding Engine.
32	Lancashire Boiler Shed.
33	Temporary 18 ft Ventil. Fan.
34	G.W.R. Co's Engineer' Office.
35	New Mission Hall.
36	T-Bob Pumping Engine House.
37	Cornish Boiler House.
38	Miner's Cabin.
39	Blacksmith's Shop.

For References Nos 1 to 21 inclusive, see Plans of Buildings, etc. at Sudbrook, Dec. 1880-1.

SCALE

100 50 0 1 2 3 4 5 6 7 8 900 FEET

PLAN SHEWING BUILDINGS &c. AT SUDBROOK & 5 M. 4 CHS. DECR 1882.

LONDON RICHARD BENTLEY & SON, 1887.

Machure & Co Lith London.

CHAPTER VIII.

FURTHER PROGRESS AND GREATER TROUBLES.

THE first nine months of the year 1883 were com- paratively uneventful months. The progress of the works was constantly increasing, and they were now in full swing throughout.

The rate of progress rose from £23,000 per month at the end of 1882, to £31,000 by mid-summer, and continued at the rate of about £30,000 per month through the remainder of the year.

The most rapid progress was made on the Gloucestershire side of the river. More shafts were sunk at intervals of 60 or 80 yards from the existing heading to the new gradient, and the whole of the bottom heading was completed down to the 'Shoots.'

The heading through the gravel which had run in, in November, 1882, was secured and safely driven to the extreme eastern end of the tunnel, where a junction was made with a small shaft sunk by the Company in 1879.

The bridges over the open cutting were commenced, and the cutting itself lowered to the right level from its eastern end for a length of rather more than a quarter of a mile, when two steam-navvies were brought upon the works, and started to take out the excavation. A shaft was sunk at a point rather more than a quarter of a mile east from the eastern tunnel face, where a public road and one of the main drains of the level crossed the railway.

At this point it had been originally intended to build an ordinary bridge ; but after the lowering of the gradient, it was found necessary to execute a short length of tunnel instead of this bridge. The shaft was sunk for the purpose of constructing this tunnel.

The invert of this tunnel was just on the top of the red marl, and the whole of the tunnel itself in gravel, with great volumes of water. Above the gravel was soft running sand, and then mud.

It was ground that required the greatest care, and could only be safely executed by being kept perfectly dry.

Two 15-inch pumps were therefore fixed in the shaft ; and before the end of the year a break-up was started in the middle of this tunnel, and about 20 yards of completed tunnel executed.

Before the end of the year rather more than a mile of full-sized tunnel was completed from the Sea-Wall Shaft. The arch was finished under the whole length of the 'Shoots,' and all but 100 yards from there to the Sudbrook Pit. The tunnel was com-

pleted for rather more than 100 yards westwards
from the Sudbrook Pit, and the bottom heading from
the same point, towards the point where the Great
Spring had broken in, was commenced.

At 5 miles 4 chains, more than 200 yards of full-
sized tunnel were put in, and about 100 yards of
arch turned.

Another break-up was also started, and the brickwork commenced in it, at about 250 yards westwards
from this shaft.

Very good progress was made from the Marsh
Pit. By the end of the year more than 200 yards
of the tunnel were completed east from this pit, and
about 700 yards westwards.

From the Hill Pit about 70 yards of tunnel were
completed east from the pit, and more than 400 yards
westwards.

The cutting on the Monmouthshire or Welsh side
of the river was commenced in March.

On the 9th February, a terrible accident happened
to some men who were working at the 5 miles
4 chains pit. The men were gathering round
the bottom of the pit at one o'clock (after midnight) to come up to supper; the one cage was
at the bottom, and four or five men had just got into
it ; and the other cage was on the upper level, from
which the skips were taken to the clay-crushing
machine, when the banksman at the ground-level (a
steady man, who had worked a considerable time at
the same employment), seeming to forget the position

of affairs, and thinking that the cage on the one side was standing on the ground-level, suddenly took hold of and pushed an iron skip right into the mouth of the pit. The iron skip, falling about 140 feet, crushed the bonnet of the cage at the bottom, and killed three of the men who were in the cage ; and the skip then rebounding among the crowd of men who stood near, killed another man and seriously injured two others.

The holes which had been stopped in the bed of the river once or twice this year required further attention, the tide washing away the clay and bags which had been placed over them ; and it may be as well to state here, that in the following year, before the tunnel was completed, when this clay was supposed to have settled as far as possible, the hole at the top was slightly enlarged and sealed with a thick layer of concrete.

The works being thus in a very advanced condition, except the length of about 200 yards adjoining where the Great Spring had broken in, it was decided to open the door in the head-wall, which had been built across the heading in December, 1880, and to take in hand the length passing the Great Spring.

At the end of May an attempt was made to open this door, but it was at once found that a quantity of rock and shale had fallen down behind it, and that it was impossible to open it. On the 30th May holes were bored through the door with augers, and a Progress of the work.

1883. piece 12 inches across was broken out. Through this a considerable quantity of soft material was forced by the pressure of the water behind, and now and then large rocks were brought down which stopped the hole. For more than two months the men continued to work, taking away material which was forced through the door by the pressure of the water behind, having continually to break up with long bars and 'jumpers' the lumps of rock which blocked the hole. At last it became evident that it would be an endless matter to attempt to work in this manner, so the bottom heading was pushed forward until it passed well beyond the point where this door was built.

A hole was then broken up from the bottom into the top heading, and all the water from behind the door allowed to pass that way.

Men then got up into the upper heading, and found that a length of 50 or 60 feet of the roof had fallen in, and that there was an enormous cavity above, but that little water was coming from that direction.

To obtain better access to the upper heading than was afforded by the hole broken up from the bottom heading, we then drove a small side heading round the end of the head-wall, and thus had double access to the heading beyond.

We commenced at once to pole and secure a heading through the mass of *débris* that had fallen from the roof. We also drove a rising heading from the end of the brickwork of the tunnel, up at an incline of about 30 degrees from the horizontal, and so bróke into the great cavern that had been left above the top of the tunnel by the falling down of the roof of the heading. Up this sloping heading we carried all the old timber that we could obtain in short lengths, and threw it forward into the cavern, hoping to fill it up, and to support the roof before further mischief occurred.

We continued to drive the bottom heading 9 feet high by 9 feet wide, and in it we built a head-wall with a door, almost directly under the door in the upper heading.

We then restored' and properly timbered the upper heading, repaired the head-wall, and hung a new door in place of the one which we had broken in our endeavours to get through We cleared down also to the top of this head-wall from the sloping heading, and built another head-wall across that to guard against accidents.

The total quantity of material forced through this door by the pressure of the water behind was 2,000 yards, showing the enormous extent of the cavity above, and the damage that had been done by not properly timbering the heading at first.

While this work was going on, on the 30th September, another serious accident occurred, happily without loss of life or injury to any of the men. A 'bond' or wire-rope, used for lifting the large cages at the Sudbrook Winding Shaft, broke about

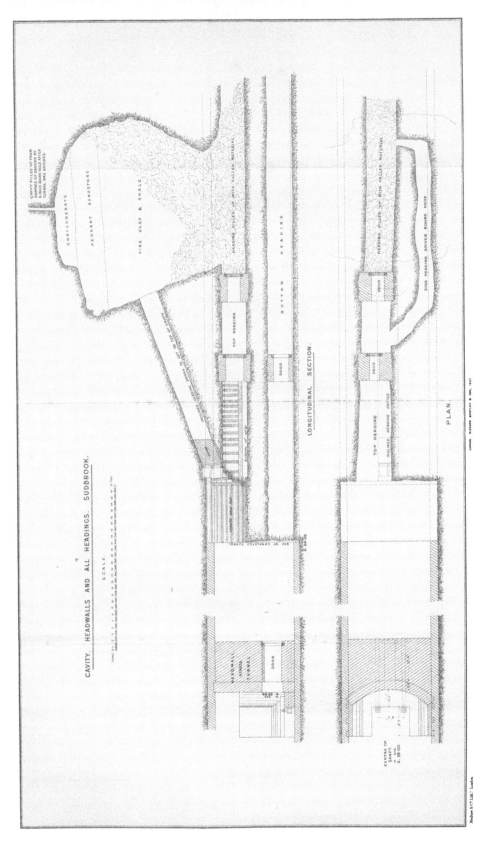

The material originally positioned here is too large for reproduction in this reissue. A PDF can be downloaded from the web address given on page iv of this book, by clicking on 'Resources Available'.

midnight. The two cages were adjusted by additional wire-ropes to balance each other, so that the
engine only had to lift the weight contained in the
cages, and not the cage itself. The bond broke just
as the one cage, containing four loaded skips of rock,
had reached the top; the other cage was at the
bottom. The first cage was not sufficiently high to
allow the banksman to turn the tumblers under it,
and when the bond broke, the upper cage being
loaded, and the other empty, the loaded cage rushed
down the pit with terrible velocity, throwing the
light cage up against the head-gearing at the
top of the pit, smashing the pulleys and the
other wire-rope; and then the light cage also
fell to the bottom, and both cages lay together
broken.

It is noteworthy that the wheels under the
loaded skips of rock which fell 200 feet into the
tunnel, being made of Hadfield's steel, were not
broken.

As the work was now so far advanced, I decided
not to replace these large cages, and fixed instead
two single cages, only large enough to carry the
cobs that were used in this part of the work for
hauling the skips. While the alterations were being
made, four of these cobs had to be stabled at the
bottom of the pit.

The following pages will show how fortunate it
was that at this very time we had taken these precautions with regard to the head-walls and doors,

and the security of the work generally between Sudbrook and the Great Spring.

Of course all possible precautions were taken at all times, but it seemed an extreme one at the moment to push forward these three head-walls, one above the other.

CHAPTER IX.

A WEEK OF TROUBLES.

On the 10th October, 1883, all the work was going Progress of the work. on favourably. Everyone connected with it was in 1883. high spirits, thinking of rapid completion ; and at six o'clock on the evening of that day, when the miners, who had worked the day-shift, fired their round of holes, no extra difficulties and no impending trouble were foreseen.

I had left the works just after six o'clock, after speaking to the foreman of the miners about the number of holes fired in the westward heading at Sudbrook just as I was leaving the office ; but I had not been at home more than an hour, when the same foreman drove up to my house to report to me that the Great Spring had broken into the tunnel in larger volume than we had ever yet met with, and that it was rapidly gaining upon the pumps.

I went down with him at once to the works, and on descending the shaft found a river 16 feet or 17 feet wide of bright clear water, flowing along the invert of the tunnel, and leaping down the old shaft into the lower workings.　As it had a drop there of more than 40 feet, the roar of the water, when you were in the tunnel, was deafening.　My first action on reaching the bottom of the shaft was to taste the water.　To my great relief I found that it was fresh, and we, therefore, had no reason to suppose that we should not be able to contend with this difficulty, as we had with so many before.

It was evident, however, that the water was coming in at a rate very much exceeding our power to pump it out, and therefore the time that would elapse before it rose through the lower workings must be spent in precautionary measures.　It was impossible to approach the head-wall in the lower heading, and the door there was open ; but men went up to see that the door in the upper heading was properly closed, and to put additional timbers in the sloping heading to secure the head-wall there.

As there was a hole existing from the tunnel at Sudbrook to the new pumping-shaft on the south side of the line, bricklayers were at once started to close this hole with brickwork in cement.　As soon as these works were started, I ascertained the following facts with regard to the inburst of the water.

The night-gang, working in the bottom heading westwards, had gone to work shortly after six o'clock, taking up skips with them, and had begun to shovel up the loose rock dislodged by the blasting, when the ganger said :

'There is more water here than usual—the

"Grip"' (a small ditch at the side of the heading) 'must be blocked. Push back a skip or two to clear it out.'

The men had hardly done so, when, to use the words of the ganger, 'the water broke in from the bottom of the face of the heading, rolling up all at once like a great horse.' It swept the men and the iron skips like so many chips out through the door and into the finished tunnel; and it was only when the water spread itself over the whole width of the tunnel that they were able to gather themselves up, and save themselves from being precipitated down the old shaft into the lower works. They were swept through the door without the power to check their passage, but they at once endeavoured to work their way back again up the heading, holding one another and clinging to the timbers at the side, to shut the door, if it were in any way possible. All their efforts failed, for the water was running down the heading in a stream 10 feet wide and 3 feet 6 inches deep, and with such rapidity and force that no man could stand against it.

Anxiously we watched the rising of the water. We found that it was rapidly gaining upon the pumps; that it was already 10 or 12 feet deep in the tunnel under the shoots; that the men had all escaped; but that the horsekeeper had, in his terror, ridden off on one of the cobs, and left three others to drown.

Finding, on my arrival at the shaft, that the

pumps were running twelve strokes per minute, I ordered them slowed down to ten. I was particularly anxious that there should be no panic.

A messenger was sent over by the last ferry-boat, to order the men on the Gloucestershire side of the river to at once commence to build a head-wall of brick in cement across the finished tunnel, west of the Sea-Wall Shaft. The men worked all the night and the next day, and completed the head-wall, leaving only a door 3 feet square at the top; but the water never rose so high as to reach the bottom of this head-wall. On the morning of the 11th the water had risen against the pumps to the height of 52 feet. On the 12th the pumps, still working steadily, held the water at 132 feet from the surface.

A council of war was held, and it being the opinion of all that the inburst of the water might be from a subterranean reservoir, which would shortly exhaust itself, and that we should only have the same quantity of water ultimately to pump that we had before the inburst occurred, it was decided to continue the pumping for two or three weeks longer. After holding it a depth of 132 feet from the surface for two days, the pumps began to gain slowly. By the 22nd they had gained 9 feet 9 inches, and by the 26th, 13 feet.

The cubical contents of the tunnel and other works filled by the water, while the pumps were continually pumping at the rate of 11,000 gallons per minute, was accurately measured; and we found

THE TEMPORARY ENGINE HOUSES, SUDBROOK.

Machure & Cº Lithº London.

THE PERMANENT ENGINE HOUSES, SUDBROOK.

LONDON RICHARD BENTLEY & SON, 1887

that the water must have run in at least at the rate
of 27,000 gallons per minute, or 16,000 gallons more
than we had pumping-power to lift.

Having thus obtained a fairly accurate measurement of the quantity of water we had to contend with, I again engaged the services of Lambert, the diver who had on a previous occasion closed the door in the long heading, as well as the use of one of Fleuss's diving dresses, in which he had previously done the work. On the 29th he tried to reach the door in the Fleuss dress, but found it impossible to do so.

I think at the time his health was bad, as he had just returned from Australia, where he had been engaged in raising a vessel, the *Austral,* in Sydney harbour.

On the 30th, however, assisted by two other divers, he went up again, dressed in his ordinary dress, and this time succeeded in closing the door.

By the 3rd November the pumps had again entirely cleared the tunnel of water, and the Great Spring was imprisoned, as it had been in January, 1881.

On the 12th October, just when the water of the Great Spring had gained its highest level and wholly drowned the works at Sudbrook and under the river, the largest of the pumps at 5 miles 4 chains broke, and in a few hours that pit also was drowned, and the works full of water.

At 7 p.m. on the 17th October, the night-shift,

consisting of about 90 men, had descended the Marsh Pit, to proceed with their work.

About 450 yards of tunnel were completed at the bottom of this pit, and two break-ups were being worked, and were in various stages of progress west of it.

It will be remembered that the gradient of the tunnel at this point rose 1 in 90 to the west.

A perfect storm of wind was blowing at the time from the south-west, and it was known that one of the highest tides of the year would occur that night, but no tide had ever been known to come so high as the works at this shaft.

Between the shaft and the river itself, in a south-westerly direction, were a number of small cottages, built by the men employed upon the works, of stone and timber; and there were also several brick cottages, owned by the firm who carried on the Tin-plate Works, and inhabited by their men.

Suddenly, in the darkness, a great tidal-wave burst over the whole of the low-lying ground between the shaft and the river. It must have come on as a solid wall of water, 5 or 6 feet high. It entered all the houses, most of which were only of one storey, and rose above the beds on which the children were asleep, The children were saved by being placed upon high tables, or even on shelves.

The bedding, blankets, and many articles of furniture, were entirely destroyed. Fortunately the houses were not thrown down.

TIDAL WAVE, MEN IN BOAT SAWING THE TIMBER.

LONDON   RICHARD BENTLEY & SON, 1867

The tidal-wave, passing beyond the houses, reached
first the boilers that worked the winding and pumping engines at the shaft, extinguishing the fires, and
then flowed down the pit with a fall of 100 feet.
There was a ladder-way from the top to the bottom
of the shaft, and by it, when the first force of the
water had passed, one or two men who were in the
bottom managed to make their escape; one unfortunate man, after climbing the ladder for about
half the height, was thrown back by the force of the
water and killed.

Eighty-three men were imprisoned in the tunnel
at the bottom of the shaft. As the water rose, they
retreated before it up the gradient.

In the darkness, and with the whole of the shaft
surrounded by water, it was extremely difficult for
the two or three who were on the top to communi-
cate with the works at Sudbrook; but at last one
man made his way through the water and gave the
alarm. The principal foreman of the works, with
his brother, and one or two of the assistant engineers
and other employés, reached the shaft, not without
difficulty—some following the line of the tramway,
wading through more than 3 feet of water; some
passing over the Great Western line, the rails of which
were, opposite the shaft, at least 6 inches under water,
and then wading through a shorter distance to the
shaft. On reaching the pit-head, where by this time
the tide was of course lower than it had been at the
first rush of the wave, everything that could be

Progress of
the work.

1883.

gathered, water-proof clothing, sacks, timber, and such-like things, were used to try to form a dam round the top of the shaft to stop back the water.

In spite of all that could be done, the water rose in the tunnel at the bottom of the shaft, to within 8 feet of the crown of the arch. Then the tide going back, and the dam at the top being more effectually made, preparations were made to rescue the men who were imprisoned below. The whole of the bottom of the tunnel and heading was under water, and the men had retreated to a stage in one of the break-ups, where they sat not knowing what their fate would be.

The men who had by this time gathered round the top of the pit were sent for a small boat, which was lowered down the pit, and launched on the water in the tunnel : a few men with lights, getting into the boat, pushed up the tunnel to rescue their comrades ; but after going a short distance they came to timbers placed across the tunnel from side to side, which blocked their progress. Returning to the shaft, they obtained a cross-cut saw, and commenced to cut away the timbers. They had been at work but a short time, when the saw dropped overboard, and they had to wait until another was procured ; but at last the men were all rescued, and brought safely to bank on the morning of the 18th.

The tidal wave had not only drowned the Marsh Pit, but had come up with sufficient force to flow

Machure & Co Lith.ᵈ London.

CUTTING, FLOODED AFTER HIGH TIDE OCTᴿ 17ᵀᴴ 1883.

LONDON. RICHARD BENTLEY & SON, 1887

in considerable quantity over the sea-wall on the
Gloucestershire side of the river, and the meadows around the shaft were, on the morning of the 18th, covered with water, which stood 8 inches over the rails of the tramway.

To the westward of the Marsh Pit the same tidal wave had flooded the whole of the meadows, and the sea bank round the cutting at the western end of the tunnel not being completed, the cutting was full of water. Fortunately, the heading from this cutting to the tunnel had not at that time been completed, or the works at the Hill Pit would also have been drowned by the action of this tidal wave.

On the morning of the 18th the works of the tunnel were in a worse position than they had been since January, 1881 ; and though each difficulty had been successfully contended with as it arose, it was yet to be ascertained whether we could hope, without great delay, to arrange sufficient pumping machinery to cope with the body of water which it was evident we had met with between Sudbrook and 5 miles 4 chains.

The pits at 5 miles 4 chains and the Marsh were soon clear of water, and work was resumed at the Marsh Pit on the 23rd, and at 5 miles 4 chains on the 19th. As soon as the pumps cleared the tunnel under the river of water, on the 3rd November, the work upon the Gloucestershire side of the river was resumed with the full force of men ; but as an additional precaution against the Great Spring, for fear of the rising heading (which had not been thoroughly secured, and was in bad ground) giving way, a large head-wall of brick in cement was built across the full-sized tunnel, 260 feet west of the Sudbrook Pit ; at the same time a concrete wall was built round the boilers, engines, and pit at the Marsh, so that even if we were visited by another wave of corresponding height, we should at least save that pit from being drowned, and the fires in the boilers from being extinguished.

The height of the tidal wave was found to be 10 feet above the calculated height for the tide on that night.

Large quantities of timber, which had been stacked at the Marsh Pit, were floated away to some distance, and one large larch-tree, 15 inches in diameter, was landed upon the top of the post-and-rail fence east of the Marsh Pit ; the fence being 4 feet 6 inches above the level of the meadows.

CHAPTER X.

THE MEANS TAKEN TO DEAL WITH THE GREAT SPRING.

It had been thought up to the time when the spring Progress of the work.

——

1883. flooded the works that we had pumping-plant upon the ground more than sufficient to deal with the greatest quantity of water that could be met with.

We had first a 75-inch Cornish beam-engine, with a 35-inch bucket-pump ; two 50-inch engines with a 26-inch pump each ; a 70-inch engine, with a 28-inch and two 18-inch pumps ; a 41-inch engine, with a 28-inch pump; a horizontal engine, with 18-inch pumps ; and nine 15-inch pumps worked by horizontal engines. It was evident that we must more than double this plant before we could hope to complete the tunnel, and then the more serious question arose, 'Where should the pumping-plant be fixed ?'

It was almost universally supposed that we should have to sink another shaft in which to place this pumping-plant.

It was advisable to obtain the plant ready made if

possible, or at any rate the engines, and inquiries were at once set on foot through the mining districts in various parts of the country to ascertain if any suitable plant could be purchased; and after many inquiries I purchased, from Messrs. Harvey and Co., of Hayle, two 70-inch beam-engines, and one 35-inch bucket-pump; two 60-inch engines, each with a 31-inch bucket-pump; and the Great-Western Railway Company ordered to be delivered at once the 37-inch plunger-pump which had been provided under the original contract, but which had not then been received.

While these inquiries were being made, and the purchases effected, I had thought of a way in which it would be possible to fix three of these pumps in the Old Pit, by closing that pit as a winding-shaft, and winding entirely from the New Pit.

The Old Pit was only 15 feet in diameter, and therefore, though it was possible to get the pump-barrels into the pit, it was not possible so to place them in the pit as to be able to take off the valve door-pieces, to examine or repair the valves; but by filling up the bottom of the shaft to within 10 feet of the tunnel level, I arranged to bring the valve-pieces into the tunnel itself, where there was plenty of room for taking off the doors and making any examinations or repairs. The accompanying drawings will show how this was effected.

At this pit we arranged to fix two 70-inch beam-engines and one 60-inch. The one 70-inch engine

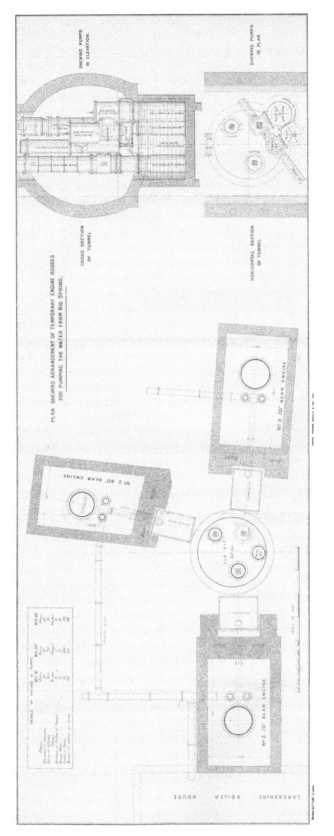

The material originally positioned here is too large for reproduction in this reissue. A PDF can be downloaded from the web address given on page iv of this book, by clicking on 'Resources Available'.

was to work the 37-inch plunger-pump, the second 70-inch to work the 35-inch bucket-pump, and the 60-inch engine to work a 31-inch bucket-pump. The second 60-inch beam-engine, with a 31-inch bucket-pump, was fixed at 5 miles 4 chains, where, in consequence of the pressure after the head-walls were closed at Sudbrook, a much greater quantity of water was coming into the works.

The first work commenced was the filling up of the bottom of the Old Pit. This was done with concrete, through which a 6-inch pipe was laid into the 9-ft. barrel, and the water which came into the pit was drained through it to the Iron Pit. By means of this pipe we were able to keep the shaft sufficiently dry, and to finish the concrete to within 10 feet of the top of the invert of the tunnel. A considerable piece of this invert had to be cut out on one side to receive the H-piece of the 37-inch plunger-pump.

The erection of four large engines of course necessitated the provision of additional boilers. These were also purchased. Ten Lancashire boilers, varying from 27 feet to 30 feet in length, and 7 feet in diameter, were purchased, and the seating of the boilers and the erection of the boiler-houses was commenced.

The estimated expense of providing this additional plant was more than £16,000, and at the time when this heavy additional expenditure was to be incurred I was already, from the unforeseen diffi-

culties we had encountered, £100,000 out of pocket.
It was, however, of no use being overcome by the
difficulties, and with as cheerful a face as possible
we set ourselves to overcome them instead.

The four pumps which were to be fixed, and one
of the 60-inch engines, had to be made at Hayle, in
Cornwall. The other three engines, which were
standing in various parts of the country—one at
Llanelly, and two others at Cornwall—had to be
taken down and forwarded by rail as quickly as
possible. A new gang under a Cornish foreman
was formed for erecting these engines.

By altering the old engine-house, where the 41-inch
engine had originally stood, and strengthening the
' bob-wall,' which carried the beam, a house was provided for one of the 70-inch engines. A house for
one 60-inch engine at 5 miles 4 chains was built of
brickwork, and for the other 60-inch and 70-inch
engines two new houses were built of stone and brick.
The accompanying drawing will show that the erection of the houses required considerable time, so that
but little progress was made before the year 1883
came to a close.

In spite, however, of these difficulties, the progress
of the other works was continually increasing, and
at the end of the year we found that, with the
arrangements that we had made, we could take
out more lengths of tunnel than we were able to
obtain bricks to line it with.

We were at that time receiving from the Catty-

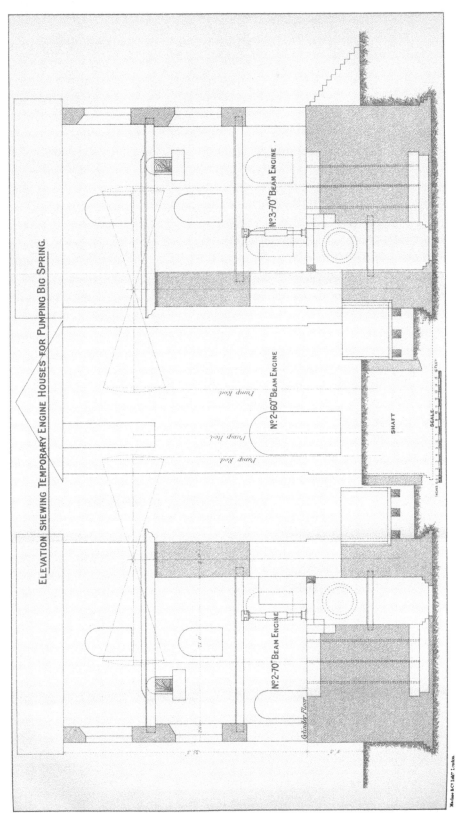

ELEVATION SHEWING TEMPORARY ENGINE HOUSES FOR PUMPING BIG SPRING.

Nº 3-70" BEAM ENGINE.

Nº 2-60" BEAM ENGINE

Nº 2-70" BEAM ENGINE

Pump Rod

Pump Rod

Pump Rod

Pump Rod

SHAFT

SCALE

Cylinder Floor

The material originally positioned here is too large for reproduction in this reissue. A PDF can be downloaded from the web address given on page iv of this book, by clicking on 'Resources Available'.

Progress of
the work.
——
1883.

brook Brick Company, near the Gloucestershire
end of the tunnel, 100,000 bricks per month.
From two makers in Staffordshire we were obtaining
500,000 bricks per month, and we were making in
our own field about 600,000 per month. The total
of 1,200,000 bricks was sufficient to complete 3,600
cubic yards of brickwork per month.

As I was very anxious that the work should be
proceeded with, and secured as rapidly as possible,
I sent an agent down to Staffordshire to secure a
larger supply, and from five different makers we
obtained contracts to supply a further quantity of
1,000,000 bricks per month.

The amount of the work done in the year 1883 is
shown upon the accompanying section.

At the end of 1883, one mile of continuous full-
sized tunnel was completed on the Gloucestershire
side of the river, and the work was rapidly progressing at ten other points on the same side from ten
'break-up' lengths. The bottom heading had been
completed throughout the whole length of the
tunnel between Sudbrook and the Gloucestershire
side.

The arch had been turned under the 'Shoots' for a
length of half a mile; the full-sized tunnel had been
completed, from the Sudbrook Shaft, for a length of
something more than a quarter of a mile. At 5 miles
4 chains shaft a length of a quarter of a mile of tunnel
and arching had been completed. At the Marsh,
rather more than half a mile of full-sized tunnel,

and at the Hill Pit rather more than a quarter of a mile of full-sized tunnel, had been completed also.

The open cuttings at each end of the tunnel were making rapid progress; four locomotives and three steam navvies being employed on the Gloucestershire side, and three locomotives with one steam navvy on the Monmouthshire side, besides large gangs of navvies filling by hand.

The 5-ft. barrel-drain, to take the water from under the 'Shoots' to the pumping-shaft, had been completed.

In the summer of this year, there was a very serious outbreak of small-pox at Chepstow. I had already (in 1881) had to deal with an outbreak of typhoid fever on the Gloucestershire side of the river, and had succeeded in stamping it out entirely by providing a fever hospital with a skilled nurse, to which all cases were removed, and then by making provision for a better supply of drinking water.

Fearing that the small-pox epidemic might very probably be brought to our works by the men who frequented Chepstow on Saturday evenings, I determined to lose no time in building a hospital for infectious diseases as near to the houses as was consistent with safety.

After consulting Dr. Lawrence (who was in medical attendance upon the men) and Dr. Bond, of Gloucester (the health officer of the district), I built a hospital, of which a sketch is given. There were

THE FEVER HOSPITAL, SUDBROOK.

LONDON. RICHARD BENTLEY & SON, 1887.

Maclure & C⁰ Lith⁰ London.

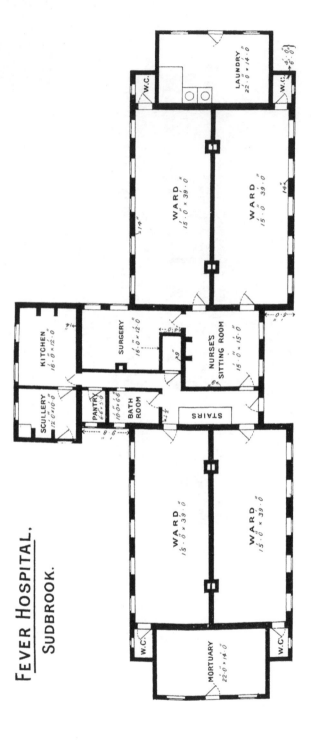

FEVER HOSPITAL.
SUDBROOK.

W.C.	
LAUNDRY 22·0 × 14·0	
W.C. 2·6 × 2·6	

WARD 15·0 × 39·0

WARD 15·0 × 39·0

14″

KITCHEN 16·0 × 12·0

SCULLERY 12·0 × 10·0

PANTRY 10·0 × 6·6

BATH ROOM

SURGERY 16·0 × 12·0

NURSE'S SITTING ROOM 15·0 × 15·0

STAIRS

WARD 15·0 × 39·0

WARD 15·0 × 39·0

MORTUARY 22·0 × 14·0

W.C.

W.C.

GROUND PLAN.

SCALE

FEET 10   5   0   10   20   30   40   50 FEET

LONDON.   RICHARD BENTLEY & SON, 1887.

four wards in the hospital, each large enough for
six beds, with ample air space.   There was also a
separate ward for delirious patients, with two beds,
and a dwelling-house and sleeping apartments for
two nurses and servants.

The building of the hospital was commenced in
July, 1883, and it was first used in the autumn of this
year.

CHAPTER XI.

SIDE-HEADING DRIVEN TO BIG SPRING—COMPLETION OF THE BRICKWORK—BIG SPRING SHUT OUT—WORKS PASSED BY BOARD OF TRADE.

Progress of the work.

1884.

THE first nine months of the year 1884 were like those of 1883—uneventful months.

The whole of the tunnel, except rather less than 300 yards, where the Great Spring had been shut back by two head-walls, was fully at work. The largest possible number of men was employed upon the works, and all the plant and machinery that could be utilized was on the ground.

The carrying out of the works had originally been entrusted to two principal foremen—Mr. John Price taking the Gloucestershire side of the river, and Mr. Joseph Talbot the Monmouthshire side, which included the work under the 'Shoots' and the work at the three outlying shafts, 5 miles 4 chains, the Marsh, and the Hill, as well as the principal shaft at Sudbrook.

The greatest number of men employed in this year was 3,628; 1,641 being employed on the

Gloucestershire side, and 1,987 on the Monmouth-

shire side.

The total length of tunnel executed on the Gloucestershire side of the river, under John Price, was 3,260 yards. In addition, he also executed about 100 yards of tunnel detached from the main tunnel, under the road known as Ableton Lane. The total length of tunnel executed under Joseph Talbot was 4,406 yards.

The length of the tunnel stated in the contract originally had been 7,942 yards; but this had been reduced in the beginning of the year 1884 to 7,666 yards; the cutting at the western end of the tunnel being lengthened to the extent of 276 yards, to provide the material required for making up the sidings at the station near Rogiet, afterwards known as Severn Tunnel Junction.

Every part of the works, except the 300 yards previously mentioned, was pushed on with rapidity, and only those ordinary accidents and delays which are inseparable from tunnel-work interfered with the progress.

Singularly enough, we generally found that when one accident occurred, others followed almost immediately. For instance, on February 15th the T-bobs of two of the pumps at 5 miles 4 chains broke down, and, in breaking, bent the crank-shaft of the engine. On the 17th the pump-rod of one of the pumps at the Marsh Pit broke; on the 18th a large pump at 5 miles 4 chains broke down; and on the 19th

the rising-main of the 35-inch pump in the Iron Pit
at Sudbrook split.

In order to relieve the water at 5 miles 4 chains,
we had for some time kept the sluices in the big
head-wall near Sudbrook partially opened ; but when
the 35-inch pump broke down it was necessary to
shut these sluices, or the works under the river
would have been drowned. Shutting the sluices
increased the pressure to such an extent at 5 miles
4 chains and at the Marsh Pit that, in consequence
of the breaking down of the pumps, these two pits
were of a necessity partially filled with water, and
the works stopped ; and it was not till March 17th
that the pumps were repaired and the work resumed
in these two pits.

The whole of the tunnel on the Gloucestershire
side, from the ' Shoots ' to the open cuttings at the
eastern end of the tunnel, was completed in August ;
and the whole of the tunnel from 5 miles 4 chains to
the western end of the tunnel was completed in
September.

Sir John Hawkshaw had determined to put down
a new pumping-shaft just outside the western end of
the tunnel, and to pump there, at a low lift, all the
rainfall from the open cutting, which, with the sea-
banks, contained an area of 14 acres. This shaft
had been sunk, and a heading through the whole of
the unfinished portion of the cutting completed, to
bring the water to the pumps fixed there.

The water from the large cutting at the eastern

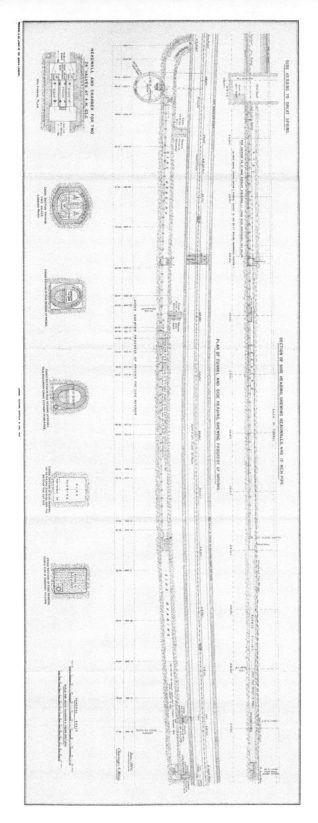

The material originally positioned here is too large for reproduction in this reissue. A PDF can be downloaded from the web address given on page iv of this book, by clicking on 'Resources Available'.

end of the tunnel was led into the tunnel itself, and pumped by the pumps fixed in the Sea-Wall pumping-shaft.

The two 70-inch engines and the two 60-inch engines were being erected in the houses provided for them, and the pumps were being fixed in the shafts.

In order to deal with the water from the Great Spring, Sir John Hawkshaw decided to drive a side-heading parallel to the centre-line of the tunnel, but about 40 feet to the north of it, from the Old Pit at Sudbrook to the point where it would intercept the spring itself.

The gradient of the tunnel rising 1 in 90, the heading was to be driven at a gradient of about 1 in 500 ; so that, when it reached the point where the Great Spring had broken in, it would be about 3 feet below the bottom of the invert of the tunnel.

The driving of this heading was commenced in July, both from the point at which it was to join the tunnel and at a point 50 yards to the west, by a cross-heading from the tunnel, in order to ensure the correctness of the lines. By the 20th September this side-heading had been driven about 18 yards past the point where the head-wall had been built across the tunnel.

Sir John Hawkshaw had further decided that, in order to reduce, if possible, the quantity of water to be dealt with—which, we knew, to a very consider-able extent came from the loose bed of the little river Neddern—to construct a concrete invert in the bed of the brook for a length of nearly 4 miles.

This work was undertaken in August, and was completed by the 7th October; and three of the large engines and pumps being ready for work, on the 30th September it was determined at once to take in hand the work between Sudbrook and 5 miles 4 chains.

The sluices in the head-wall were gradually opened, and at the same time a hole was broken through the brickwork at the top of the head-wall.

In the first day's work the head of water behind the head-wall was reduced 4 feet 6 inches, and after three days' pumping all the water was out of the heading, and the foremen were able to go up and inspect the point where the spring had broken in. All the work was found in good condition, having been properly secured by timber just before the 10th October, 1883.

Another cross-heading was driven from the tunnel from behind the head-wall to connect with the side-heading; and the side-heading was pushed forward, and on the 19th December reached a large open joint in the strata, which had formed the channel for the subterranean water. A drawing is given showing a plan of this fissure, and the manner in which it was tapped by the side-heading.

By diverting the water of the spring into the side-heading the tunnel itself was left almost perfectly dry, except where, in taking out the invert, we

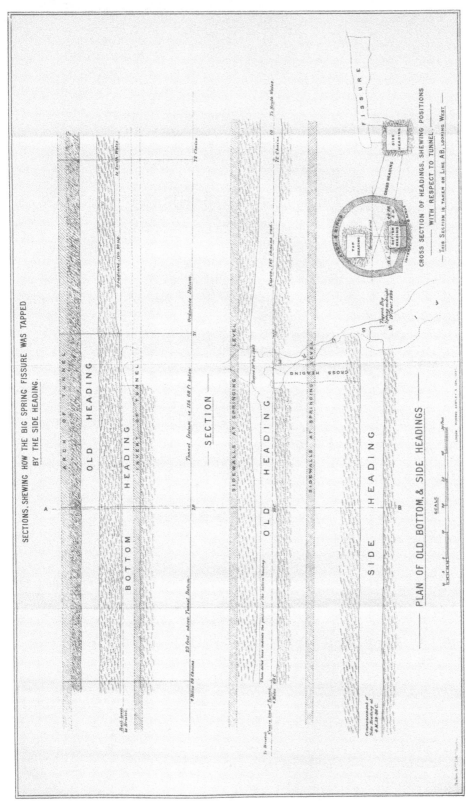

The material originally positioned here is too large for reproduction in this reissue. A PDF can be downloaded from the web address given on page iv of this book, by clicking on 'Resources Available'.

crossed other fissures in the rock in which the water stood a foot or two higher than in the one tapped by the side-heading.

As a precautionary measure, before starting the pumps at the end of September, we had built across the side-heading opposite the great head-wall in the tunnel a head-wall of brickwork in cement, with a door hung as in the previous head-walls. By the 6th October we had so completely mastered the water from the Great Spring that we found it possible to stop one of the largest of the pumps, and hold that in reserve. On the 17th October we effected a junction between the top heading from 5 miles 4 chains and the heading from Sudbrook, making a complete passage from end to end of the tunnel ; and as it happened that, without warning, Sir Daniel Gooch, the Chairman of the Great Western Company, and the Earl of Bessborough, one of the directors, came upon the works the same day, we were able to pass them, at two o'clock on the 17th, through the last link of the works, for it was now possible to walk through from the open cutting on the Gloucestershire side to the open cutting on the Monmouthshire side. Though it was possible to pass through, it must not be supposed that the road was the very best or cleanest imaginable. The journey could be made by riding on trolleys from the eastern end of the tunnel to the shaft at Sudbrook ; and could also be made from 5 miles 4 chains to the western end of the tunnel in the same way ;

but between Sudbrook and 5 miles 4 chains there
was not only a quantity of water dripping from the
roof, but there were ladders to climb, and, at any rate
on the 17th October, a small and not very clean
hole to crawl through, with a face of 4 feet or 5 feet
between the two headings to be climbed up.

On the 9th August of this year the members of
the Institution of Mechanical Engineers, who were
holding their annual meeting at Cardiff, visited the
Severn Tunnel Works, and were able to examine
them all with the exception of the short length at
the Great Spring. Arriving by train from Cardiff
at the point since known as Severn Tunnel Junction,
they were transferred from passenger carriages to
trucks fitted up with seats, and drawn by one of the
locomotives employed upon the works, in which
they passed through the open cutting to the mouth
of the tunnel. Descending from the trucks, some
members of the Institution elected to walk over the
top of the tunnel, but the greater number continued
their examination, walking through the finished
works underneath. Meeting with a rough road
to travel, and here and there some little water, other
members chose to go up by the cages at the Hill
Pit ; others, again, dropped off when the Marsh Pit
was reached ; but a goodly number continued their
walk through the tunnel to 5 miles 4 chains, and
were then drawn to the surface, and found their
train waiting for them. After examining the brick-
making machinery they proceeded to Sudbrook,

PLAN SHEWING BUILDINGS &c. AT SUDBROOK & 5 M. 4 CHS. DECR. 1884.

SCALE

LONDON. RICHARD BENTLEY & SON, 1887.

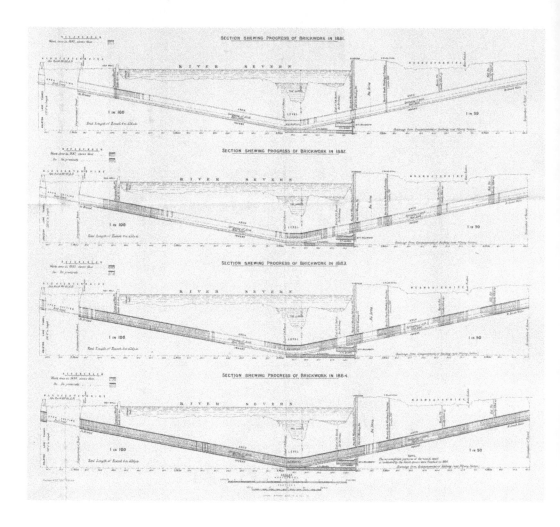

The material originally positioned here is too large for reproduction in this reissue. A PDF can be downloaded from the web address given on page iv of this book, by clicking on 'Resources Available'.

made a thorough examination of all the pumping Progress of the work.

1884.
and ventilating machinery, descended the Sudbrook
Shaft, and examined the tunnel under the river;
and then, returning to the surface, wound up with
what Englishmen can never dispense with—a good
lunch in the school-room; and on leaving the works
expressed themselves highly delighted with all that
they had seen.

On the 15th September the members of the
South Wales Institute of Engineers also visited the
works, and made a thorough examination of them.

On the 22nd October the laying of the rails of
the permanent road was commenced from the
Gloucestershire side of the river; and on the 20th
November the largest of all the pumps, the 37-inch
plunger, was completed, and started to work in the
Old Pit, thus giving us two pumps in reserve over
and above what was necessary to deal with the
Great Spring.

The bottom heading, between Sudbrook and
5 miles 4 chains, was completed on the 16th December, two days under five years from the date on
which the contract had been signed.

The progress section, which is given on the other
side, shows the work that was done in the year 1884;
but shortly stated, the whole of the tunnel was
completed, except about 200 yards between 5 miles
4 chains and Sudbrook, and 500 yards of invert
under the 'Shoots,' where the connections were to
be made with the 9-ft. barrel and the 5-ft. drain.

The work remaining to be done being now
confined to a very short length, it was necessary to
reduce largely the number of men employed, but as
many were retained as could be kept at work, and
in order to hasten completion, four break-ups were
started in the short length near the Great Spring, at
distances of only 44 yards apart. The running
lengths were also commenced from both faces, and
thus ten faces were kept at work through this short
length of tunnel.

On opening out the full-sized tunnel, the fissure
through which the Great Spring had passed was
found to follow a most erratic course. In one place
it passed directly across the tunnel from side to side,
nearly at right angles to the centre line of the work.
At another place it passed from side to side in an
oblique direction, running for some small distance
directly under one of the side walls. At another
point where the tunnel had been perfectly dry, while
the mining was done, the lifting of almost the last
stone out of the invert set free an immense body of
water which no pumps underground could cope
with. At another point the water boiled up from a
hole 18 feet in depth under the invert with such
force that stones, the size of a man's fist, dropped
into the water would descend about 10 feet, and then
begin to flutter like a leaf in the wind, and be
thrown out again by the water. Into this hole a
cast-iron pipe was lowered, attached to a bend at
the top to lead the water into the side-heading.

The cavity round the pipe was closed with concrete
and brickwork in cement, and the place made perfectly sound and tight.

Where the fissure crossed the tunnel at right
angles, two 15-inch cast-iron pipes were laid side by
side under the invert, and were then covered with
concrete and brickwork in cement, and an extra
thickness of brickwork in cement was put into the
tunnel invert.

The point where the water rose in the invert of
the tunnel by the lifting of what I have called the
last stone gave more serious trouble.

At last, at the suggestion of Sir John Hawkshaw,
a wall was built round the hole, and inside it the
water was allowed to rise. When it had risen about
4 feet, it remained stationary, as at that height the
water escaped by other openings into the side-heading. When the water was stationary the whole of
the cavity below the invert was filled up with Port-
land-cement concrete lowered down a wooden tube.
After this had been allowed time to set, a hand-
pump was set to work in the top of the hole, and it
was found that the concrete had made a perfectly
water-tight joint, and the hole was easily pumped
dry. The concrete had been kept down to a level
of 3 feet below the top of the invert, and on this the
invert of the tunnel, 3 feet thick of brickwork in
cement, was built, without any water to interfere
with the work; and on April 18, 1885, at 8 a.m., the
last length of the brickwork of the tunnel was keyed

in, at a point nearly midway between Sudbrook
and 5 miles 4 chains shaft.

The brickwork of the tunnel having been completed, the side-heading was arched over above the
level of the water from the Great Spring, and over
the arch filled up with rubble masonry from the
point where the western end intersected the fissure.
When the door was reached, one of the sluices was
shut on July 13, and a 12-inch pipe being laid to
carry the water from the other, the rest of the heading was filled up with rubble masonry, and the
opening through the side-wall of the tunnel closed.
The 12-inch cast-iron pipe was laid to bring water
from the spring to the boilers, and a small shaft was
sunk at the side of the tunnel, nearly over the
end of the side-heading, and a pipe led down
through the shaft to connect with the pipe in the
heading.

In addition to the sluice at the head-wall, a sluice
was fixed on this pipe at its eastern end. The
sluice by the head-wall was opened before the heading was filled with masonry, and the sluice at the
other end was kept at our command for supplying
water as might be hereafter required. At the end
of the cast-iron pipes a pressure-gauge was fixed
to the pipes to indicate the pressure of the water,
and so tell how much it had risen day by day. The
following table shows the rate at which the water
rose :

Progress of the work. 1885.

TABLE SHOWING RATE AT WHICH PRESSURE ROSE AFTER SLUICES WERE SHUT AUGUST 11, 1885.

Date.	Pressure in lbs.	Date.	Pressure in lbs.	Date.	Pressure in lbs.	Date.	Pressure in lbs.
Aug. 11	nil	Aug. 26	39¾	Sept. 14	47½	Oct. 27	54
" 12	9	" 28	40	" 15	47¾	" 31	54½
" 13	12	" 29	41½	" 17	48	Nov. 4	55
" 14	17	" 30	42½	" 18	48¼	" 7	55½
" 15	18¾	" 31	43	" 20	48¾	" 9	56
" 16	21	Sept. 1	43½	" 22	49	Dec. 3	57
" 17	22½	" 2	43¾	" 25	49¼	" 7	57¼
" 18	25	" 3	44½	Oct. 1	49½	" 20	57¼
" 19	27	" 4	45¼	" 4	49¾	" 21	30*
" 20	29¼	" 5	45½	" 7	50	" 22	25
" 21	32½	" 6	45¾	" 10	50¾	" 23	16
" 22	34¾	" 7	46¼	" 11	51	" 29	13
" 23	36½	" 8	46¾	" 15	52½	" 30	nil†
" 24	37½	" 9	47	" 17	53		
" 25	38¾	" 13	47¼	" 24	53½		

* Valve opened 4″.
† Valve fully opened.

N.B.—Pressure is given in lbs. per square inch.

On 5th September Sir Daniel Gooch, accompanied by Lady Gooch and a party of friends, came with a Great Western engine and passenger train to inspect and pass over the works.

Arriving at Severn Tunnel Junction at 11 o'clock, the train passed through the open cutting and through the tunnel into the open cutting on the Gloucestershire side; returning again to the Monmouthshire open cutting, the train passed over the line laid to bring materials to the works to Sudbrook, where two or three hours were spent pleasantly in inspecting the engine-house, school-rooms, men's houses, mission-hall, etc.

On the day the train passed through, the water had risen in the ground outside the tunnel to a height of about 105 feet, and the pressure shown upon the pressure-gauges was 45½ lbs. per square inch. At this pressure a small quantity of water found its way through the joints in the brickwork, and in some places made quite a respectable shower into the tunnel.

On the 7th September I left the works for South America. There had been a great strain upon me for many months, and I was glad to get on board the royal mail steamboat *Neva*, at Southampton, on the 9th, and to think that the tunnel was all right, and that I should be out of reach of all telegrams and letters for at least four weeks. We reached Buenos Ayres on the 6th October, and I purposed remaining there till the 17th November, to verify information which had been given to me, and to complete my estimates for the Madero Port at that place. On the 30th I received a telegram: 'Sir John Hawkshaw says you must come home on the 1st.' I replied that it was impossible. I finally left South America on the 18th November for Southampton.

Arriving at Southampton on the 14th December, I received the first intimation that there had been further troubles at the tunnel. After I had left on the 9th September, the water, continuing to rise

J. CLARKE HAWKSHAW.

ENGRAVED BY W.H.GIBBS FROM A PHOTOGRAPH BY WITCOMB.

London. Richard Bentley & Son: 1888

in the ground where the Great Spring had been
stopped back, had been of course producing a greater and greater pressure upon the brickwork, till the pressure had at last risen to $57\frac{1}{4}$ lbs. on the square inch. Under that pressure the bricks in the tunnel began to break, pieces flying off them with reports like pistol-shots, and the water shooting through the broken bricks quite across the tunnel.

Two of the largest of the pumps which had been fixed to deal with the Great Spring in the Old Shaft had been taken out before I left for South America, and only the 31-inch pump worked by a 60-inch beam-engine was left in that pit to deal with any extraordinary quantity of water which might find its way through the brickwork.

Mr. J. Clarke Hawkshaw had gone to South America with me, and had remained there when I returned. Sir John Hawkshaw was naturally extremely anxious about the state of the works, and when I had seen him and heard what had occurred since I left, I visited the works, and found them in a very serious, if not dangerous, condition, and I at once asked Sir John's permission to take off the pressure. He was of opinion that our pumping power was not sufficient, and that the water from the spring was coming in behind the work in greater volumes than we were able to pump.

Having, however, obtained his sanction to making the attempt, I broke out the pipe adjoining the sluice valve at the eastern end of the side-heading, and opened the sluice gradually as the pumps were able to take the water. The sluice was opened on 21st December, and the pressure was rapidly reduced from $57\frac{1}{4}$ lbs. on the square inch to 30 lbs.

Sir John Hawkshaw then determined that it was necessary to provide pumping-plant to pump the whole of the water from the spring, and not to subject the brickwork of the tunnel to the enormous pressure it would have to sustain to exclude this water.

Arrangements were made with me to sink a large shaft at the side of the tunnel, 29 feet in internal diameter. In this shaft were to be fixed six large pumps with six 70-inch Cornish beam-engines fixed in a house which entirely covered the shaft. It was also determined to fix two 65-inch engines with two new pumps in the pumping-shaft at 5 miles 4 chains, and two 41-inch beam-engines with 29-inch pumps in the shaft at Sea-Wall.

A Guibal Fan, 40 feet in diameter and 12 feet wide, was ordered by the Company, and the designs made for the necessary buildings, comprising fan-house, engine-house, and boiler-house. Two Lancashire boilers, each 26 feet in length by 7 feet in diameter, were provided for the fan, and a space provided for a third boiler, which was afterwards added. Twelve Lancashire boilers 28 feet in length, and 7 feet in diameter, were provided for the pumping engines at Sudbrook, and a new engine and boiler-house were built at 5 miles 4 chains, in which

four Cornish boilers already there belonging to the
Company were fixed with three new Lancashire
boilers.

The 29-ft. shaft, which was 35 feet in external diameter outside the brickwork, the circular wall being 3 feet in thickness, was sunk in the manner described in the account of the sinking of the other shafts, and without any difficulty from water, as a bore-hole was dropped down into the side-heading below.

The shaft was commenced the 8th February, 1886, and completed to the bottom on the 7th April. The brickwork lining of the shaft was completed on the 3rd June. The building of the engine-house at the top of the shaft was commenced on the 9th February, and completed on July 8th, 1886. The engines in this case were not only constructed, but were erected by Messrs. Harvey and Co., of Hayle, the pumps being fixed by me ; and the first engine and pump were ready to start and were started on the 1st July, 1886.

The length of the 12-inch cast-iron pipe laid in the side-heading was 515 feet ; and when this pipe was running full bore, it still required a head of 70 feet to force the whole of the water through the length of pipe.

On the 1st July we had been able for some weeks to take the whole of the water this pipe would bring. The second of the large engines and pumps was ready to start on the 4th August. In the meantime,

the rubble masonry with which the side-heading had
been filled had all been removed, and the heading
inside had been lined, as far as possible, with brickwork in cement. In order to complete the brickwork, a cross-heading from the tunnel was opened
into the side-heading at a point west of the head-
wall, the water being temporarily diverted through
that cross-heading ; the brickwork of the invert was
completed, and the pipes broken off at the back of
the sluice in the head-wall. The water was then
allowed to flow down the brick-lined heading to the
big pumps ; and the length of 12-inch pipes being
thus reduced to 12 feet, the sluice there was opened
till the pressure was reduced to 12 lbs. on the square
inch.

The second sluice (for there were two) through
this head-wall was then opened, and all the water
taken from the spring ; and the pressure being
entirely taken off the brickwork, all broken bricks
were carefully cut out, and all damaged brickwork
repaired.

On the 20th September the third of the large
engines and pumps was started to work.

The fan was also completed on the 31st August,
and the line was opened for goods traffic on the
1st September, 1886.

On the 17th November the tunnel-works were
inspected by Colonel Rich, the Government Inspector, with a view to the opening for passenger
traffic.

Colonel Rich, who is noted for the great attention he pays to all details, made a most exhaustive inspection of the tunnel itself and of all the machinery provided and fixed, and expressed himself perfectly satisfied with all that had been done : a copy of his report is appended below.

The work was finally opened for passenger traffic on the 1st December, 1886, nearly fourteen years from the time the Great Western Company had first commenced the works, and as nearly as possible seven years from the time they had let the contract to me.

[COLONEL RICH'S REPORT.]

Railway Department,
Board of Trade,
22nd Nov., 1886.

SIR,

I have the honour to report, for the informa-tion of the Board of Trade, that, in compliance with the instructions contained in your minute of the 10th inst., I have inspected the Severn Tunnel Railway, which connects the Great Western South Wales Union Railway with the South Wales Railway.

The new line is 8 miles 26 chains long.　The gauge is 4 feet $8\frac{1}{2}$ inches.

The ruling gradient is 1 in 90, and the sharpest curve has a radius of 10 chains.

This curve is at the junction with the South Wales Railway, and is only about $1\frac{1}{2}$ chains long.

The railway is double throughout, except about 37·54 chains at Pilning, which will be partly doubled before the line is opened, and the rest as soon as

Progress of the work.

—

1886.

the alteration of the South Wales Union Railway
is completed.

The permanent way consists for about 4 miles
29 chains in the tunnel of a longitudinal sleeper road,
with a bridge-rail that weighs 68 lbs. per lineal
yard ; and the remainder of the Great Western
standard pattern sleeper road with an 86-lb. bull-
headed rail and 43-lb chair.

The rails are made of steel ; the railway is well
ballasted and well fenced.

The only stations are at Pilning and at Severn
Tunnel Junction.

The works consist of three bridges under the
railway, which have wrought-iron girders on brick
abutments ; three brick bridges and one wooden
bridge over the railway ; eight culverts ; and the
tunnel under the Severn, which is 7,664 yards long.

The tunnel is lined with vitrified brick in cement.

The sides, which rise about 7 feet above the rail-
level, carry a semicircular arch of 26 feet diameter,
which varies in thickness from 27 inches to 36 inches;
and there is a brick invert, 18 inches to 3 feet thick,
throughout the tunnel, which has a brick semicircular
drain about 3 feet 6 inches wide over the centre,
and a brick barrel-drain 5 feet in diameter under the
invert. This latter carries the surface-water to a
shaft at the north end of the tunnel, and drains the
cuttings at both sides of the railway.

In addition to these there is an old heading 12 feet
wide at the north end of the tunnel, which has been
utilized for drainage purposes.

There are ventilating shafts at each side of the

river. The one at the south side is 15 feet diameter ;
and the one at Sudbrook, which is at the north side, is 18 feet diameter.

The ventilating fan, which is at the north side of the river, is 40 feet diameter and 12 feet wide. It can be worked at a velocity of 60 revolutions per minute : less than half this speed is estimated to be sufficient to ventilate the tunnel.

I enclose herewith a statement of the several pumping shafts, and of the pumps that work through them to drain the tunnel, which, in addition to the water from the cutting at each side, and some very slight leaks through the brickwork of the tunnel, has intersected an underground stream of a considerable size. The water from this stream has to be pumped into the Severn from a shaft 29 feet diameter, which has been constructed to catch this water.

There are arrangements for shutting off and diverting the stream water by means of sluices from the 29-feet diameter shaft, and also for intercepting the drainage water from the cuttings and pumping it up from shafts at each side of the Severn.

In addition to the two ventilating shafts there are five pumping shafts with a power capable of pumping 38 million gallons of water per diem. The maximum amount of water to be pumped has been 30 million gallons, and the minimum has been 23 million gallons per diem up to the present time, so that there is an excess of pumping power of 8 million gallons per diem ; but when the whole of the permanent pumping and engine power, which

is now erecting, is completed, there will be a pumping power equal to 66 million gallons per diem, which
will give 36 million gallons per diem in excess of the
maximum quantity of water that has to be raised
from the tunnel and discharged into the Severn.

The works appear to be very good and substantial, and to have been carried out with great
care and judgment.

This work was commenced twelve years since,
and the actual work of building the tunnel was
commenced in 1881.

The difficulties of dealing with the large quantity
of water, and particularly of dealing with the underground stream, which runs at a great velocity, have
been considerable, but have now been successfully
overcome, and the result is a tunnel of unusually
large dimensions, which is particularly dry. The
top of the tunnel is about 145 feet under the level
of high-water spring tide, and about 50 feet under
the bed of the river at its deepest point.

The means of ventilation are ample, but did not
act well when I made my inspection.

This only requires a little attention and regulating.
I was informed that no inconvenience has been felt
from a want of ventilation during the running of
the numerous goods and coal trains that have been
sent through the tunnel for some time past.

I have, etc.,

(Signed)   F. H. RICH,

Colonel R.E.

The Secretary,
    Railway Department,
        Board of Trade.

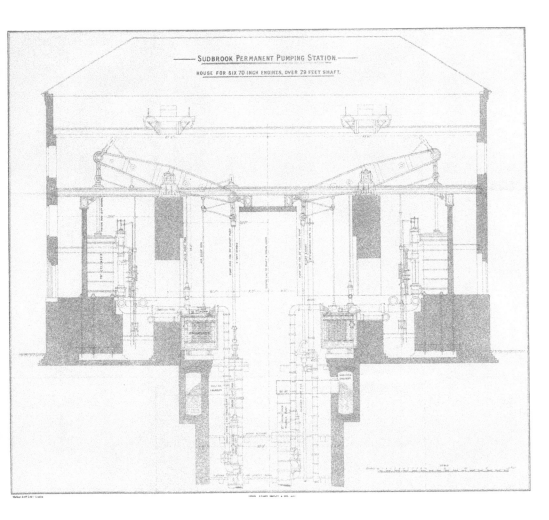

The material originally positioned here is too large for reproduction in this reissue. A PDF can be downloaded from the web address given on page iv of this book, by clicking on 'Resources Available'.

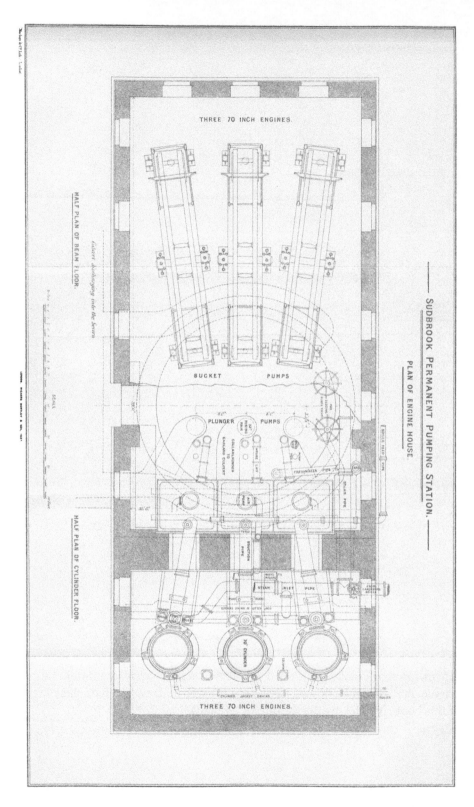

The material originally positioned here is too large for reproduction in this reissue. A PDF can be downloaded from the web address given on page iv of this book, by clicking on 'Resources Available'.

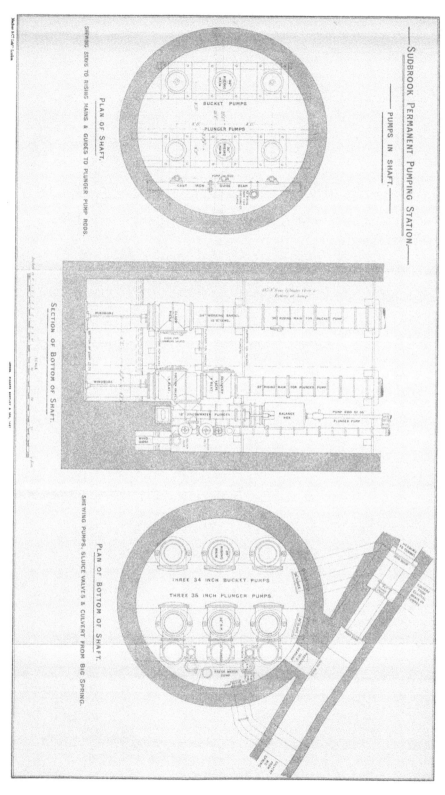

The material originally positioned here is too large for reproduction in this reissue. A PDF can be downloaded from the web address given on page iv of this book, by clicking on 'Resources Available'.

CHAPTER XII.

THE MEN BY WHOM THE WORK WAS DONE.

AT the head of these must be placed, of course, the chief engineer, Sir John Hawkshaw.

It would be presumptuous of me to say much about Sir John himself, or to speak of his vast experience, and the name he had made in connection with the great works he had carried out.

I had the honour of working under him on the East London Railway, continuing Brunel's original Thames Tunnel under the London Docks, through Wapping, Shadwell, Whitechapel, to join the Great Eastern Railway—a work where the difficulties met with were second only to those encountered in the Severn Tunnel.

I also carried out for Sir John and Mr. J. Wolfe Barry, in 1883 and 1884, the completion of the inner circle of the metropolitan railways in London, by the construction of the City lines from the Mansion House Station to the Tower.

Born in 1811, Sir John was seventy-five years of age when the tunnel was completed.

He was constant in his visits to the works, and when there made the most minute and particular inspection of every part of them, riding in a skip-trolley, or walking through the headings, dressed in miner's costume, and keeping the most wonderful run of all the details of every particular part of the works in his head.

His son and partner, Mr. J. Clarke Hawkshaw, took, under Sir John, the principal charge of the works. His visits were of course more frequent than Sir John's, and every ordinary question with reference to the works was brought before him by the resident engineer.

Mr. Harrison Hayter, the third partner in the firm of Hawkshaw, Son, and Hayter, arranged all the details of the contract when it was first entered into ; and during Mr. Clarke Hawkshaw's absence in South America from the beginning of September, 1885, to the end of January, 1886, Mr. Hayter had, at a very difficult time, to enter upon the charge of the works, and to work out the details of the new pumping machinery and steam-power to be provided to pump the Great Spring.

The originator of the scheme for constructing a tunnel under the Severn, Mr. Charles Richardson, was born on the 29th August, 1814. He was the third son of Mr. Richard Richardson, of Capenhurst Hall, in Cheshire, a well-known landowner in those parts, and Deputy-Lieutenant for the county, and Chairman of the Quarter Sessions at Knutsford.

HARRISON HAYTER.

ENGRAVED BY W.H. GIBBS FROM A PHOTOGRAPH BY WINDOW & GROVE

London·Richard Bentley & Son,1888.

He died in 1820. Mr. Charles Richardson, after leaving Edinburgh University, became a pupil of Mr. Isambard Kingdom Brunel in 1834. He set out a good part of the Great Western Railway at the Bristol end, and put down the trial shafts on the Box Tunnel; but his first engagement on a public work was in the Thames Tunnel, under Sir M. I. Brunel, the father of Mr. I. K. Brunel.

In 1838 he undertook the supervision and setting out of the line from Swindon to Cirencester, and was afterwards resident on the Stroud Valley Railway, where he built his first equilibrated arch from his own design, under a trackway of 1 in $2\frac{9}{10}$.

In 1846 he carried out the line between Hereford, Ross and Gloucester, and afterwards made the Bristol and South Wales Union Railway, which latter was the introduction to the Severn Tunnel Railway.

M. A. G. Luke was the resident engineer under Messrs. Hawkshaw, Son, and Hayter. Mr. Luke was at the works from 1st January, 1881, up to their entire completion. He had to superintend the whole of the setting out, to keep all the accounts of the work done, and to make innumerable designs for new works required to meet emergencies. I can say that, in a long experience on public works, I have never known a resident engineer more cautious or painstaking, more anxious to protect the interests of the Company, and at the same time pleasant to transact business with.

Several younger engineers were employed under Mr. Luke as assistants; but perhaps the most remarkable character on Sir John's staff was his chief inspector, Mr. Isaac Jackson.

Mr. Jackson had been with Sir John Hawkshaw for a great many years, and Sir John had great confidence in him. A thoroughly practical man he was, able to go into every detail, and to keep every department in proper working order, the terror of any of the men who tried to deceive him. He was now and then known to give way to a little temper, when it was as well to give him as wide a berth as possible.

He had been inspector for Sir John Hawkshaw on the East London Railway, through all the most difficult part of the work; and after ten years' knowledge of him, I believe that no man could be found of sounder practical judgment in his department than Mr. Jackson—willing to assist in any proper way, fertile in expedients, but determined to have things done properly, and his orders carried out to the letter.

Mr. Jackson had at one time eight other inspectors working under his orders, so that the execution of the work was carefully looked after.

To come to my own staff. This contract, being an exceptionally heavy one, was not entrusted to the charge of an agent with more or less independent authority; for, considering the magnitude and the anxieties of the work, I had determined to live

as near to it as possible, and to keep the principal charge of it in my own hands.

My first lieutenant was Mr. F. R. Kenway. At the time the tunnel works were commenced by me, Mr. Kenway had been with me seven years on several railway contracts, the last of which were the East London and the Dover and Deal railways.

The post of first lieutenant at the Severn Tunnel was something like the post of first lieutenant on board a line-of-battle ship, with a good deal of responsibility, and a vast amount of detail to attend to ; but as the other heads of departments, engineering machinery and works were men thoroughly competent to carry out their own parts, this labour was considerably lightened to him.

Mr. A. O. Schenk was chief of my engineering staff, and was responsible for all the setting out, a work which he performed in a manner most highly creditable to him. Giving the lines in a tunnel is always a difficult operation, and requires great care. In the Severn Tunnel the difficulties were much greater than in an ordinary tunnel.

Mr. Schenk had been a pupil of Sir William Armstrong's, and was therefore a thoroughly mechanical, as well as civil engineer. His assistance in the mechanical department was often of the greatest value, and his application of compressed air to the pumps for lowering the heading proved a perfect success; and some of the setting out was really extraordinary.

I have stated before that there were about 1,500 'lengths' in the whole tunnel, and under the river the tunnel is straight for about 2¾ miles; and though these separate lengths were put in from more than forty different 'break-ups,' connected by small headings, it is impossible to detect any deviation from the straight line upon the work that has been executed.

At 5.4 the pumping-shaft, about 45 feet to the south of the tunnel, was built from the bottom, from a centre point given by Mr. Schenk, which was set out from the tunnel through a small heading, and proved to be perfectly true with the shaft as set out from above; but the following was the most extraordinary piece of 'setting out' executed by Mr. Schenk: The original pumping-shaft at Sudbrook had been tubbed with iron, and three large pumps were fixed in it. When the 5-feet barrel-drain was completed to the new shaft, a heading on a sharp irregular curve was driven, and the bottom of the Iron Pit was excavated, with a roof of about 10 feet of rock between it and the bottom of the pumps, the pumps being constantly at work. A careful survey was made from the 5-feet barrel to find the centre of the Iron Pit. A point was fixed by Mr. Schenk, the lower part of the shaft built; and when a hole was broken through the roof into the Iron Pit, the point given did not vary one inch from the true centre-line of the shaft.

Mr. J. H. Simpson, who was chief of the mechan-

ical department, had been with me for about fourteen years before the Severn Tunnel contract was
entered into. Thoroughly careful and painstaking,
it was amusing to try to pin Mr. Simpson to a
promise to finish anything by a given date. He
certainly had not the same opinion of the value of
time as men of his generation are supposed to have.
' Slow but sure ' was his motto, and not a bad motto
either on such works.

The principal foreman on the Monmouthshire
side was Joseph Talbot.

Joe came to me from some works abroad in the
year 1865. He carried out a large part of the
Metropolitan District Railway, doing the work from
Gloucester Road Station to South Kensington, and
afterwards doubling the South Kensington Station.
He also did the works in front of Somerset House
on the Thames Embankment. He was for some
time on docks, then did the heaviest part of the
East London Railway, and afterwards the tunnel on
the Dover and Deal Line.

He was born on a tunnel on the South-Eastern
Railway, not far from Dover, and has been engaged on tunnel-work all his life up to the present
time. His father was a good miner before him, and
five of his brothers also followed the same occupa-
tion.

Thoroughly at home in all that a miner had to do,
the only complaint he ever made was against the
hardness of the ground. He once gave it as his

deliberate judgment that no tunnels ought to be made in hard ground ; they ought to be made only where the ground was soft.

Joe, of course, was *first* in exploring everything. He was the first to go through the Shields into the western heading, the day before he took Mr. Clarke Hawkshaw and me up the same heading. He was first up the long heading when the water was out, and brought me back a *mile in the dark*, when our lights went out on my first trip up, saying, ' Put your hand on my shoulder ; I can go along all right in the dark.' He was first up the heading after the ' panic,' and though he could see nothing then to alarm him, took the wise precaution of shutting the door in the head-wall before he returned. He generally was the one chosen to act as guide to strangers visiting the works, and I think there are very few visitors to the underground workings who will not remember ' Joe ;' and I have no doubt he often has a quiet laugh to himself when he recalls the adventures of some of them. On one occasion two ladies expressed a wish to go through a certain part of the workings.

' You'll find it *rather* wet,' said Joe.

' Oh ! we do not mind that,' they said ; ' we have come prepared to get wet.' And so they had, as far as regards water falling from the roof, being equipped in miners' donkey-jackets and sou'-wester hats ; but they little thought they would have *to wade through two feet of water for some dis-*

tance ! which they did pluckily, rather than turn back.

On the occasion of the visit of the mechanical engineers, Joe had quietly arranged that they should not go away without some idea of what blasting was like underground. A favourable spot being chosen in one of the 'break-ups,' about fifty holes were drilled, charged, and pinned all ready for firing at a given signal. The miners, being in the secret, were all quietly at work while the visitors were filing past; but as soon as the last of them had reached the safety point, each miner lit his fuse and ran, and the result was as successful as Joe could have wished. The fuses having been cut in various lengths, so as to give the effect of a bombardment, and the tunnel having been completed except just at this particular 'break-up,' the sound echoed and re-echoed along it to such an extent as to startle the most experienced of the underground travellers; and I think some of the younger ones were not sorry when the next shaft was reached, and they could get into the open air again, and get the ringing sound out of their ears.

The principal foreman on the Gloucestershire side was John Price. Price always said he was not a Welshman, but a Shropshire man. However, I think that matter is a little doubtful. He had also been at work through the Metropolitan and the Metropolitan District Railways, and was on the East

London Railway when it was opened in 1876.   The
line was nearly ready for the Government inspection, and I was walking over it with Price, when he
tripped over a board that lay upon the permanent
way, and falling with his knee across the edge of a
sleeper, split his knee-cap, which left him slightly
lame ever since.

Whether from this or what other reason I do not
know, he took to reading more than men of his
stamp generally do, and being experienced in
tunnelling and intelligent, he pushed his way forward more rapidly than others, and was able to take
higher positions.

He carried out his part of the work in the most
efficient manner, and with great rapidity, so that in
November, 1884, I was able to remove him to
another contract before the whole of the works of
the Severn Tunnel were completed.

Second in position on the works on the Monmouthshire side was Joe Talbot's brother, Mat
Talbot, William Cox occupying a similar position on
the Gloucestershire side.

Under these men, who were called 'walking
gangers,' there were employed, when the works were
in full swing, as many as fifty gangers on the Monmouthshire side, and seventeen on the Gloucestershire side ; each ganger having under him, on the
average, five miners and twenty-one labourers ; and
in addition to these there were the men called 'runners out,' who pushed the full skips from the head-

ings to the point where the ponies or the wire rope

were attached to them.

The greatest number of men at work at one time was 3,628.

In almost all parts of the tunnel the men, when at work, had to wear either waterproof clothing or flannels. These clothes were provided by me, and large rooms were erected at each pit-top where the men could change their clothes, and at each room there was a man on duty day and night to see to the safety of the men's own clothes, and to superintend the drying of the wet garments.

The average amount per week earned by the miners, taking into account all lost time, was £1 18s., and by the labourers £1 7s. 6d.

All the work except the brickwork was done 'day-work,' *i.e.*, the men were paid by the day or hour.

The brickwork was done 'piece-work,' the tunnel being divided between two sub-contractors, Mr. Stephen Morse doing that portion which was under J. Talbot, and Mr. Edward Silverton that portion which was under J. Price.

At one time there were employed in taking account of the time made by the men, seven timekeepers by day, and five by night; and at the same period seven pay-clerks were employed, whose duty it was to make up the pay-books, and pay the men every Saturday. The pay was made at seven different places, in order to divide the men into gangs of reasonable size, and to save them from walking long distances for their money.

The men who were on the night-shift, which on Saturdays commenced at 2 o'clock, were paid first, from 12 o'clock to 1 o'clock, after which the day-shift were paid.

As many as thirteen assistants were employed in the engineering department, and thirteen clerks in the accountant's office.

The amount paid in wages to day-work men in the largest pay, on Dec. 21st, 1884, was £4,372 13s. 9d. There were about 3,100 men on the pay-book.

				Per hour.		10 hours per day.
Foremen	...	...	14	10d.	...	8s. 4d.
,,	...	...	54	9½d.	...	7s. 11d.
,,	...	...	9	9d.	...	7s. 6d.
			31	8½d.	...	7s. 1d.
Skilled labour :			275	8d.	...	6s. 8d.
Carpenters, Miners,			88	7½d.	...	6s. 3d.
Fitters, Engine-			183	7d.	...	5s. 10d.
drivers, and Tim-			166	6½d.	...	5s. 5d.
bermen			456	6d.	...	5s. 0d.
			441	5½d.	...	4s. 7d.
Labourers	...	...	761	5d.	...	4s. 2d.
,,	...	...	46	4¾d.	...	4s. 0d.
,,	...	...	252	4½d.	...	3s. 9d.
,,	...	...	47	4d.	...	3s. 4d.
Boys	...	...	54	3½d.	...	2s. 11d.
,,	...	...	79	3d.	...	2s. 6d.
,,	...	...	82	2½d.	...	2s. 1d.
,,	...	...	29	2d.	...	1s. 8d.

In addition to the day-work men, about 500 bricklayers and labourers were employed during the same week upon the brickwork, which was done 'piece-work,' as before stated.

No less than 76,400,000 bricks were used in the

construction of the tunnel and bridges.

These bricks were vitrified bricks, from the Cattybrook Brick Company, near New Passage, on the Bristol and South Wales Union Railway; from the Fishponds and Bedminster Company, near Bristol; from Staffordshire; and from our own brickyard, near the Five-mile-four-chain Shaft.

The quantity from each is as follows:

Cattybrook	-	-	- 19,125,440
Fishponds	-	-	- 7,229,100
Staffordshire	-	-	- 21,944,460
Our own	-	-	- 28,101,100

The quantity of Portland cement used on the works was 36,794 tons, the whole of which was brought from the Medway or the Thames, some by water to Newport, some by water direct to a wharf at the tunnel, but the greater part of it came by rail from Brentford, to which place it was carried in barges up the Thames.

The tonite used for blasting purposes amounted to about 250 tons, and several magazines had to be erected for the safe storage of this and other explosives.

The minimum quantity of water pumped when dealing with the Big Spring was 23 million gallons daily; the maximum quantity 30 millions. For more than a year the average quantity pumped daily was 24 million gallons.

To give some idea of this immense quantity of water : It is sufficient to supply a town about the size of Liverpool or Manchester, and is about one-sixth of the quantity daily consumed in London. In one year it would form a lake about 1,000 acres in extent and 10 yards deep. The total pumping power provided—66 million gallons per day, about half the supply of London – would form in one year a lake nearly 3,000 acres in extent and 10 yards deep.

All the water pumped from the Severn Tunnel during the time it was under construction would form a lake about 3 miles square and 10 yards deep.

The first passenger-train from London to South Wales passed through the Tunnel on July 1, 1887. Not the slightest hitch has occurred in the working, and the speed of the trains has often been as much as a mile a minute. The ventilation has been perfect, and the relief to the traffic between South Wales and the South-West of England has been very sensibly felt. No doubt this traffic—with the accommodation afforded by the Tunnel—will be greatly increased.

A considerable amount of the Severn Tunnel traffic for the South of England passes over the main line between Bath and Bristol, and thence on to the Salisbury line, which joins the main line at Bathampton. To provide for this traffic large sidings are being constructed near Bristol, and between Bristol and Bathampton a number of refuge

sidings have lately been made, into which a goods Progress of the work.
train can be shunted quickly. A few refuge sidings 1886.
have also been made between Bathampton and
Trowbridge.

Sir Daniel Gooch, when speaking at the general
meeting of the Great Western Railway Company,
held at Paddington, on February 10th, 1888, said :
'With regard to the Severn Tunnel, many of you
gentlemen have passed through it, and it is going on
as well as we can possibly expect. In fact, the
traffic has now become so large that it is more than
we can handle, particularly the goods and coal
traffic, and we are at the present moment making
arrangements for putting the block system in the
Tunnel itself, so as to divide it into lengths. The
distance between the two blocks is now 8 miles ; we
are dividing that into three intermediate blocks to
enable us to get the traffic through it. We keep no
separate account of the traffic which passes through
the Tunnel, as that would be a troublesome and
expensive operation. But I am glad to say that the
tendency is to show that the traffic through the
Tunnel is increasing very largely indeed. We have
no difficulty with the Tunnel ; it is working exceedingly well.'

No surer testimony than this, or from a higher
authority, could be given of the success of the
Severn Tunnel.

APPENDIX.

DESCRIPTION OF THE LARGE PUMPS BY MESSRS. HARVEY,
OF HAYLE FOUNDRY.

Six engines, with a stroke of 10 feet in the cylinder, of the type known as the Single-Acting Cornish Beam Engine, are erected in one house—three on either side of a pit, which is 29 feet in diameter, and 180 feet deep. Those on one side are connected to three pumps of the plunger type ; those on the other, to three drawing-lifts, or bucket-pumps.

These engines, as well as all the other permanent engines at this station, were erected by Messrs. Harvey and Co., Limited, of Hayle, Cornwall ; as the six engines are similar in every respect, and represent very fairly the type of engine adopted throughout, a description of one will serve for the whole.

All parts of the engines are duplicates ; thus any one piston-rod or valve will serve for either engine.

The pumps, valve-boxes, and valves of either type, are also duplicates.

Only a few spare parts, therefore, are required in case any unforeseen accident should disable one of the engines.

Cylinder. The cylinder is of 70 inches internal diameter, by 12 feet long. It is securely fixed in an outer case, leaving a space of $1\frac{1}{2}$ inches between the two, which, when working, is filled with steam direct from the boiler, thus preventing undue condensation in the working cylinder. The condensed steam from this case is conducted through the drain-pipes back again to the boilers.

The cylinder and case are mounted on a strong cast-iron bottom, and attached thereto by 36-1⅜ inch bolts. **Cylinder Bottom.**

The cylinder bottom has cast on it the aperture to which the exhaust nozzle is attached, and is secured by five long hold-down bolts to a massive foundation of masonry 15 feet in depth.

The piston is of the ordinary metallic kind, having a deep spring ring, packed at the back with rubber or gasket, and secured by a joint ring ground in true on piston and ring. **Piston.**

The rod, of wrought-iron, is connected to the piston by a deep cone, and secured by a wrought-iron cotter and fore-lock.

The cylinder-cover is of cast-iron, fitted with a deep stuffing-box, in which provision is made for a circulation of steam at boiler pressure, which prevents the possibility of air leaking through into the cylinder. **Cylinder Cover.**

The cover is also fitted with an outer case of polished cast-iron, and with suitable arrangements for lubricating the piston and cylinder.

The steam, in its passage through the cylinder, is regulated by four valves, three of which—the governor, steam, and equilibrium—are fixed in the top nozzle, and the fourth—the exhaust valve—is fixed in the botton nozzle. **Nozzles and Valves.**

All these valves are of the Cornish double-beat pattern, and are cast of gun-metal.

The governor valve is entirely under the control of the engine-driver, who can regulate the flow of steam according to any distinct variation of the pressure in the boiler.

The steam, equilibrium and exhaust valves are worked by the Cornish valve-gear.

The engine load being constant, no variation is required in the expansion, and the point of cut-off is therefore definitely fixed at about one-fourth of the length of stroke. Provision is made for cutting off the steam at from one-eighth to three-quarters of the stroke if required.

The top and bottom nozzles are connected by two vertical polished pipes, one of which is used as a steam-pipe, and the other conducts the steam—on the opening of the equilibrium valve—from top to bottom of piston. This pipe is fitted with a throttle-valve (under the control of the driver) which regulates the speed of the engine on the return or outdoor stroke.

Cataract. The number of strokes performed by the engine is controlled by the cataract, which automatically regulates it from one to about fifteen strokes per minute.

By an ingenious contrivance the speed of the six engines is regulated by two cataracts, which, although allowing each engine to perform the same number of strokes per minute, so regulates them that no two engines make their stroke at the same moment; this insures uniform pressure of steam, and an even discharge of water into the culverts. Provision is made in case either of the engines remains idle.

Engine Beam The engine-beam is of wrought-iron throughout, being built up of plates and angles having a web of strong section, and made of double-faggoted iron, secured to, and entirely surrounding, the outer edge. This web alone is sufficiently strong to resist the entire load. The centres are all accurately bored to receive the respective pins. The main pin, or gudgeon, is secured by two keys placed at the quarters, and rests in two massive gun-metal bearings, supported by heavy cast-iron stools and wall-plate. The wall-plate is secured to the main wall of building.

Parallel Motion. A parallel motion of polished wrought-iron is fitted to each end of the beam, which parallelizes the path of piston and pump-rods, the anchorage of same being attached to strong wrought-iron spring beams and girders extending across the building, and secured to the wall.

The piston and pump-rods are attached to these motions by strong wrought-iron caps.

Condenser. The condenser is of the surface-condensing type, consisting of about 400 galvanized wrought-iron tubes, secured by screwed ends into wrought-iron plates above and below.

These plates are bolted on to a cast-iron head and bottom-piece, the latter also carrying the air-pump, which is of the ordinary type.

The whole is contained in a large cast-iron cistern, supplied with a continuous stream of cold water by a pump worked from the main pump-rod.

The exhaust steam passes from the cylinder through the tubes, and being condensed by the action of the cold water without, collects at the bottom of the condenser, and is pumped back into the boilers.

The pumps are of two kinds, three being of the bucket-lift type and three of the plunger, all with a stroke of 9 feet. **Pumps.**

The bucket-lift has a working barrel of 35 inches internal diameter, with a gun-metal lining throughout, and fitted with a gun-metal bucket, carrying Husband's patent four-beat valve on same. The bucket is made extra deep, and without packing. **Bucket-Pump.**

It is connected with cotter and cap connections to the wrought-iron rods, $6\frac{1}{2}$ inches diameter.

The bottom valve is also of gun-metal of the four-beat type, seated in a strong cast-iron valve-box, and secured thereto by a suitable steel cross-head. A large circular door is provided for the ready removal of valves. The valve-box rests on a strong cast-iron windbore, having from two to three thousand $1\frac{1}{4}$ inch holes in it, the whole being supported by, and secured to, massive foundation-plates at the bottom of well.

The rising-main is of cast-iron of 39 inches diameter, in sections 9 feet long, having a wrought-iron collar-launder at the head for the delivery of the water into the culverts. The pump-head is provided with a suitable guide for the pump-rods.

Each plunger-lift has a plunger-pole, 35 inches diameter, accurately turned, and working into a pole-case, having a suitable stuffing-box with gun-metal bushings. **Plunger Lift.**

This plunger-pole is surmounted with a large load-box, which is filled with cast-iron weights, which, together with the rods, balance the column of water in the rising **Plunger.**

main. The pump-rods are of wrought-iron similar to those in the bucket-lift, and are guided at four equidistant points in the shaft by suitable guides, secured to strong iron girders built into the walls of the pit.

Valves. The suction and delivery valves are of gun-metal and of the four-beat type, as before described, seated in strong cast-iron valve-boxes, having windbores and bed-plates similar to those of the bucket-lift.

The pole-cases are mounted on, and secured to, a strong wrought-iron box-girder filled with concrete, and built into the walls of the pit.

The rising-main is of cast-iron 32 inches diameter, and in sections 9 feet long.

The flanges are all accurately faced, and those of the valve-boxes and pole-cases have a strong wrought-iron ring shrunk on them as an extra precaution against sudden shocks.

All pumps and valve-boxes, with valves in place, were tested by hydraulic pressure on the contractor's works to upwards of four times the working-load.

Fresh-water Pumps. To supply the condensers with cold water, there are six 12-inch house-lifts in the 29 feet pit, worked by a set-off from the iron pump-rods, one to each engine. These are of the plunger type, and take their supply from the respective pump-heads of the rising-mains, and deliver same into the several condensing cisterns.

There are also two 12-inch fresh-water pumps in the same pit. They are worked off the 35-inch plunger-lifts by a set-off placed immediately over the plunger-pole, and they take their supply from a fresh-water spring, cut at or about the bottom of the well.

They are also of the plunger type, having gun-metal four-beat valves and cast-iron rising-mains, delivering fresh-water into a reservoir at the surface to supply the neighbourhood.

The pit is amply provided with various ladders and stages, affording easy access to all parts of the pump-work. The valve-boxes being all about the same level, and the most important part of the system, a platform is placed there, on which are stored the spare valves.

A simple contrivance is also provided by which the valves can be changed very rapidly.

The engine-house is traversed in its entire length by a powerful traveller, capable of handling any part of the engine or pumps.

Steam is supplied to the engines from a battery of twelve double furnace Lancashire boilers placed in an adjoining building. **Boilers.**

Two main steam-pipes connect the boilers with the engines. These pipes are arranged with sluice-valves, so that the engines can be worked singly or all together.

The temporary pumps used during the construction of the works are as follows: **Temporary Pumps.**

At the Iron Pit, Sudbrook, two 26-inch plungers (the Bulls), with 10-foot stroke, raising 231 gallons per stroke each, or 462 the two together.

Also one 35-inch bucket-pump, with 9-foot stroke, raising 376 gallons per stroke.

At the new (12-foot) Pit at Sudbrook, a 28-inch plunger-pump, with 10-foot stroke, raising 267 gallons per stroke; also two 18-inch plunger pumps, with 10-foot stroke, raising 110 gallons per stroke each, or 220 gallons the two together.

At the Old (15-foot) Pit at Sudbrook, erected specially to pump the Great Spring, a 37-inch plunger-pump, with 10-foot stroke, raising 467 gallons per stroke; also a 35-inch bucket, with 9-foot stroke, raising 376 gallons per stroke; and a 31-inch bucket, with 9-foot stroke, raising 298 gallons per stroke.

At Five-mile-four-chain Pit, one 31-inch bucket-pump, 9-foot stroke, raising 298 gallons per stroke; one 28-inch bucket, 9-foot stroke, raising 240 gallons per stroke; two 18-inch plungers, 8-foot stroke, raising 88 gallons each per stroke, or 176 gallons the two; also a 5-inch and an 8-inch plunger, the two raising 25 gallons per stroke.

At the Marsh Pit, two 15½-inch plunger-pumps, with 7-foot stroke, the two raising 114 gallons per stroke, and a 15-inch bucket, raising 54 gallons per stroke.

At the Hill Pit, two 15½-inch plunger-pumps, the two raising 114 gallons per stroke.

At Benacre Bridge, two 15-inch bucket-pumps, raising 108 gallons the two.

Taking the average number of strokes per minute as ten, these pumps represent 36,000 gallons per minute. The actual delivery will be about 20 per cent. less than the theoretical, so that these pumps would raise about 27,000 gallons per minute.

On the Gloucestershire side, at Sea-wall Shaft, there were two 15½-inch plungers and two 15-inch bucket-pumps, raising in all 224 gallons per stroke.

At Green Lane, for the open cutting, two 15-inch bucket-pumps, raising 108 gallons per stroke.

At Ableton Lane, for the bridge, the same.

The total number of pumps provided during construction was thirty, and these were capable of raising 44 million gallons per day.

Four of these pumps—the 37-inch plunger, the 35-inch bucket, and the two 26-inch plungers (the Bulls)—are now permanent pumps.

Permanent Pumps. The permanent pumps provided for the Great Spring and the drainage of the tunnel and cuttings are:

At the Iron Pit, Sudbrook—two 26-inch plungers, with 10-foot stroke, raising 231 gallons per stroke each, the steam cylinders being 50 inches diameter, stroke same as pump; horse-power per stroke from water raised 13·2 per engine, or 15·9 from indicator; ordinary strokes per minute 7, maximum 12; lift for water 190 feet. Also one 35-inch bucket-pump, with 9-foot stroke, raising 376 gallons per stroke; the steam cylinder 75 inches diameter, stroke 10 feet; beam engine, horse-power per stroke from water raised 21·7, from indicator 25·7; ordinary strokes per minute 10, maximum 12; lift 190 feet.

In the 12-foot Shaft, Sudbrook, one 37-inch plunger-pump, with 10-foot stroke, raising 467 gallons per stroke ; the steam cylinder 70 inches diameter, stroke 10 feet ; beam engine, horse-power per stroke from water raised 27, from indicator 32·5 ; ordinary strokes per minute 6½, maximum 9 ; lift 192 feet.

These three pumps pump tunnel water only—the drainage from the 5-foot culvert—and in dry weather one of them working at the ordinary rate is sufficient.

At the 29-foot Shaft, Sudbrook, three 35-inch plungers, with 9-foot stroke, each raising 376 gallons per stroke (the three 1,128 gallons), and three 34-inch plungers, each raising 355 gallons per stroke (the three 1,065 gallons)—the engines have already been described ; steam cylinders 70 inches diameter, 10-foot stroke ; horse-power per stroke from water raised, plunger-pump 20, bucket 19 ; ordinary strokes per minute 7 and 8½, maximum 10 and 12 ; lift 167 feet. In dry weather only three of these pumps work at one time ; one plunger-pump, at least, is among the number, on account of the fresh-water lift.

At 5 miles 4 chains, one 35-inch plunger, with 9-foot stroke, raising 376 gallons per stroke, and one 34-inch bucket, 9-foot stroke, raising 355 gallons per stroke ; the steam cylinders 65 inches diameter, 10-foot stroke ; beam engine, horse-power from water lifted 15·30 and 14·4 ; ordinary strokes per minute 7½, maximum plunger 10 and bucket 12 ; lift 134 feet. Generally only one pump works at a time.

At Benacre, two 20-inch plungers, with 6-foot stroke, raising 82 gallons per stroke, or 164 gallons the two ; the steam cylinders 22 inches diameter, stroke same as pump ; Bull engines, horse-power per stroke from water lifted 1·32 ; ordinary strokes per minute 9, maximum 18 ; lift 56 feet. These pumps drain the cutting, and do not work in summer, the water being raised from 5 miles 4 chains.

On the Gloucestershire side, at Sea-wall Shaft, one 29-inch plunger and one 29-inch bucket, 9-foot stroke, each raising 258 gallons per stroke ; steam cylinders 41 inches diameter, 10-foot stroke; beam engines, horse-power per stroke from water lifted 7·66 each pump; ordi-nary strokes per minute average 5¼. maximum plunger 10, bucket 12; lift 98 feet. These pumps take water from cutting and a length of tunnel; as a rule, one only works at a time.

Where the horse-power from water raised only is given, there were no indicators on the engines. The indicated horse-power will be about 25 per cent. more than the actual horse-power.

The deliveries of the pumps given above are theoretical. The actual deliveries will be about 20 per cent. less. The pumps, therefore, are capable of raising about 66 million gallons per day, as stated in Colonel Rich's report.

THE END.

BILLING AND SONS, PRINTERS, GUILDFORD.
S. & H:

Milton Keynes UK
Ingram Content Group UK Ltd.
UKHW032139181024
449640UK00018B/243